SpringerBriefs in Electrical and Computer Engineering

Control, Automation and Robotics

W0017356

Series Editors

Tamer Başar
Antonio Bicchi
Miroslav Krstic

For further volumes:
http://www.springer.com/series/10198

Andrew P. White · Guoming Zhu
Jongeun Choi

Linear Parameter-Varying Control for Engineering Applications

 Springer

Andrew P. White
Guoming Zhu
Jongeun Choi
Department of Mechanical Engineering
Michigan State University
East Lansing
USA

ISSN 2192-6786 ISSN 2192-6794 (electronic)
ISBN 978-1-4471-5039-8 ISBN 978-1-4471-5040-4 (eBook)
DOI 10.1007/978-1-4471-5040-4
Springer London Heidelberg New York Dordrecht

Library of Congress Control Number: 2013934097

Printed on acid-free paper

Springer is part of Springer Science+Business Media (www.springer.com)

To our patient, kind, and loving families

Preface

The objective of this brief is to carefully illustrate a procedure of applying linear parameter varying (LPV) control to a class of dynamic systems via a systematic synthesis of gain-scheduling controllers with guaranteed stability and performance. The existing LPV control theories rely on the use of either $\mathscr{H}_\infty$ or $\mathscr{H}_2$ norm to specify the performance of the LPV system. The challenge that arises with LPV control for engineers is twofold. First, there is no systematic procedure in applying existing LPV control system theory to solve practical engineering problems from modeling to control design. Second, there exists no LPV control synthesis theory to design LPV controllers with hard constraints. For example, physical systems usually have hard constraints on their required performance outputs along with their sensors and actuators. Furthermore, the $\mathscr{H}_\infty$ and $\mathscr{H}_2$ performance criteria cannot provide hard constraints on system outputs. As a result, engineers in industry could find it difficult to utilize the current LPV methods in practical applications. To address these challenges, in this brief, gain-scheduling control with engineering applications is covered in detail, including the LPV modeling, the control problem formulation, and the LPV system performance specification. In addition, a new performance specification is considered which is capable of providing LPV control design with hard constraints on system outputs. The LPV design and control synthesis procedures in this brief are illustrated though an engine air-to-fuel ratio control system, an engine variable valve timing control system, and an LPV control design example with hard constraints. After reading this brief, the reader will be able to apply a collection of LPV control synthesis techniques to design gain-scheduling controllers for their own engineering applications. This brief provides detailed step-by-step LPV modeling and control design strategies along with a new performance specification so that engineers can apply state-of-the-art LPV control synthesis to solve their own engineering problems. In addition, this brief should serve as a bridge between the $\mathscr{H}_\infty$ and $\mathscr{H}_2$ control theory and the real-world application of gain-scheduling control.

The material presented in this brief is the result of research performed to develop gain-scheduling controllers using LPV control theory. Our goal at the beginning of this research was to develop a systematic procedure for designing gain-scheduling controllers. Since we started working in this area, we have written numerous journal and conference publications to disseminate our work.

Specifically, material from the journal papers [60–62] make up a large portion of this brief. In addition to the material from these three journal papers, we have also included a portion of our recent research on designing gain-scheduling controllers that can provide hard constraints on system outputs.

The intended audience of this brief are control engineers who are interested in designing gain-scheduling controllers for practical problems. The examples included in this brief will provide them with insight and guidance when designing gain-scheduling controllers using LPV methods for their practical problems. Control research engineers are also expected to be able to use this brief. Finally, this brief is also capable of being used as a teaching supplement to introduce graduate students with a prerequisite understanding of robust control to the area of LPV control.

We would like to acknowledge our co-authors Dr. Ryozo Nagamune and Dr. Zhen Ren for their contributions to the papers they helped us publish. Specifically, we would like to thank Dr. Ryozo Nagamune, from the University of British Columbia, for his valuable contributions to our paper "Gain-Scheduling Control of Port-Fuel-Injection Processes" during the revision process. We would also like to thank Dr. Zhen Ren for his hard work developing and building the test bench for the variable valve timing actuator. Additionally, we would also like to thank Dr. Xiaojian Yang for his work developing the mixed mean value and crank-based engine model used to validate the gain-scheduling controller developed in Chap. 4 of this brief.

East Lansing, MI, January 2013   Andrew P. White
                                                                                    Guoming Zhu
                                                                                    Jongeun Choi

Contents

Acronyms

A/F	Air-to-fuel ratio
BMEP	Brake mean effective pressure
CAN	Controller area network
DI	Direct injection
ECU	Engine control unit
EGR	Exhaust gas recirculation
HIL	Hardware-in-the-loop
IC	Internal combustion
LFT	Linear fractional transformation
LPV	Linear parameter varying
LMI	Linear matrix inequality
LTI	Linear time invariant
LTV	Linear time varying
MAP	Manifold air pressure
OCC	Output covariance constraint
PFI	Port fuel injection
PI	Proportional-integral
PID	Proportional-integral-derivative
PWM	Pulse width modulation
SISO	Single-input single-output
UEGO	Universal exhaust gas oxygen
VVT	Variable valve timing

Chapter 1
Introduction

1.1 Motivation

The goal of the research that this book is based on has been to establish a systematic procedure for the design of gain-scheduling controllers. In industry, gain-scheduled controllers are normally developed with long hours of ad-hoc tuning and calibration through, for example, engine dynamometer and vehicle field tests. While these controllers are often used successfully in many practical applications, the design process through which they are obtained is less than ideal. Not only is the process expensive and time consuming, but more importantly it may not guarantee the stability and performance of the closed-loop system for all possible time-varying parameters. In addition, the performance of the closed-loop system with gain-scheduling controllers designed in this way is dependent on the experience of person doing the calibration. In order to meet the challenges posed by the strict requirements facing many industries these days, a systematic process for designing gain-scheduled controllers with guaranteed performance and stability for all time-varying parameters is needed.

One promising solution is the advanced control theory known as linear-parameter varying (LPV) control [1–3, 11–14, 24, 45, 53, 65, 66, 67, 70]. LPV systems are time-varying systems whose time-varying components consist of measurable parameters that can vary over time. Over the years, many developments have been made in the area of LPV control theory. In the beginning, LPV control theory mainly consisted of heuristic approaches that were carried over from classical gain-scheduling control, and as such these controllers provided no guaranteed stability, robustness, or performance. The authors of [53] provided analysis conditions for these heuristic approaches that can produce guaranteed stability with slowly varying parameters. Thankfully, more advanced methods based on the convex optimization of linear matrix inequalities (LMI) have been developed [1–3, 11–14, 24, 45, 65, 66, 67, 70].

Initially, the small gain theorem was applied to LPV plants with linear fractional transformational (LFT) dependence on the time-varying parameters [3, 45]. This approach allowed the parameter variations to be complex (i.e. have both real and imaginary parts). However, since the time-varying parameters in LPV systems

A. P. White et al., *Linear Parameter-Varying Control for Engineering Applications*, SpringerBriefs in Control, Automation and Robotics, DOI: 10.1007/978-1-4471-5040-4_1, © The Author(s) 2013

rarely have imaginary parts, this was considered a major source of conservatism in gain-scheduling controller design using this method. Due to this, another method was developed that used a single or parameter-dependent quadratic Lyapunov function in the analysis and control design for LPV plants [2]. However, since this method allowed for arbitrarily fast parameter variation, it can produce conservative results with slowly varying parameters. To handle this problem, known bounds on the rate of parameter variation were incorporated into the analysis conditions by [1, 67, 70]. However, the method used by [1, 67, 70] formulates the control synthesis problem as a semi-infinite convex optimization with parameter-dependent LMIs, which requires gridding of the parameter space to provide a finite number of LMIs over which a convex optimization can be performed. A unified scheme was developed in [66] that joins both the small gain theorem and Lyapunov function approaches in an effort to provide a flexible approach for control engineers to trade-off between performance improvement, controller complexity, and design effort. However, to incorporate known bounds on the rate of parameter variation, the method developed in [66] still requires gridding of the parameter space, which increases the complexity of implementing the controller in practice. In [65], parameter dependent Lyapunov functions were used to develop LPV control synthesis conditions with a finite number of LMIs for continuous-time LPV systems with LFT parameter dependency. The authors of [65] also include the problem formulation for discrete-time LPV systems with LFT parameter dependency, but the controller formula is not provided for the discrete-time case, which means that gridding of the parameter space is still necessary. An alternative method which does not require gridding of the parameter space was provided by [11–14, 24] for affine-parameter dependent Lyapunov functions.

Although there has been a considerable amount of research on the design of gain-scheduling controllers via LPV control theory, there is still room for improvement. All of the LPV methods previously mentioned specify the performance of the LPV system as either $\mathcal{H}_\infty$ or $\mathcal{H}_2$ performance. Normally, when these performance criteria are used, some sort of weighting scheme must be used as well. In the case of $\mathcal{H}_\infty$ performance, frequency dependent weights are generally selected either to model the input and output signals (see Chap. 4) or to shape certain closed-loop transfer functions (see Chap. 5). For $\mathcal{H}_2$ performance, usually input and output weighting matrices are selected by the control designer to specify a linear quadratic cost function to be minimized by the $\mathcal{H}_2$ controller synthesis. The difficulty that arises with these methods is that real (unweighted) system performance is not easily related to the weighted $\mathcal{H}_\infty$ and $\mathcal{H}_2$ performance criteria. Furthermore, the $\mathcal{H}_\infty$ and $\mathcal{H}_2$ performance criteria, as will be discussed in more detail in Chap. 3, cannot provide hard constraints on system outputs.

The open question is, how do we bridge the gap between practices used in industry and the advanced practices used in academia? The answer is to develop LPV controller synthesis methods that allow the use of physically meaningful design constraints. By considering the ℓ_2 to ℓ_∞ gain performance of the closed-loop system, physically meaningful performance design constraints with hard bounds can be defined. This addition to the current LPV control theory is expected to be very useful for engineers working on practical applications in industry.

1.2 Book Overview

1.2.1 Providing Hard Constraints for Gain-Scheduling Controllers

This book considers the optimal control of polytopic, discrete-time LPV systems with a guaranteed ℓ_2 to ℓ_∞ gain. Additionally, to guarantee robust stability of the closed-loop system under parameter variations, $\mathcal{H}_\infty$ performance criterion is also considered as well. Controllers with a guaranteed ℓ_2 to ℓ_∞ gain and a guaranteed $\mathcal{H}_\infty$ performance (ℓ_2 to ℓ_2 gain) are mixed $\mathcal{H}_2/\mathcal{H}_\infty$ controllers. Normally, $\mathcal{H}_2$ controllers are obtained by considering a quadratic cost function that balances the output performance with the control input needed to achieve that performance. However, to obtain a controller with a guaranteed ℓ_2 to ℓ_∞ gain (closely related to the physical performance constraint), the cost function used in the $\mathcal{H}_2$ control synthesis minimizes the control input subject to maximal singular-value performance constraints on the output. This problem can be efficiently solved by a convex optimization with LMI constraints. A major contribution of this book is the characterization of control synthesis LMIs used to obtain a state-feedback LPV controller with a guaranteed ℓ_2 to ℓ_∞ gain and $\mathcal{H}_\infty$ performance. A numerical example is presented to demonstrate the effectiveness of the proposed LPV method.

1.2.2 Application of Gain-Scheduling Control

The contribution of the book also lies on illustrating how to apply such advanced LPV control synthesis techniques to practical applications in a step-by-step manner. In particular, the methods reviewed in Chap. 2, have been applied to real control problems encountered in the control of internal combustion engines. Specifically, a gain-scheduling controller design for the air-to-fuel ratio control of engine port-fuel-injection processes is presented in Chap. 4. These methods were also used to develop the observer-based mixed $\mathcal{H}_2/\mathcal{H}_\infty$ output-feedback controller for the hydraulic variable valve timing actuator in Chap. 5.

An event-based sampled discrete-time linear system representing a port-fuel-injection process based on wall-wetting dynamics is obtained and formulated as an LPV system. The system parameters used in the engine fuel system model are engine speed, temperature, and load. These system parameters can be measured in real-time through physical or virtual sensors. A gain-scheduling controller for the obtained LPV system is then designed based on the numerically efficient convex optimization or LMI technique. To demonstrate the effectiveness of the proposed scheme, both simulation and hardware-in-the-loop (HIL) simulation results are presented. The HIL simulations not only show that the designed gain-scheduling controller is effective on a complex mixed mean-value and crank-based engine model [68], but it also demonstrates feasibility of implementing the designed gain-scheduling controller on actual hardware that could be used to control an engine.

For the hydraulic cam phasing actuator, a family of linear models previously obtained from a series of closed-loop system identification tests [48, 49] is used to design a dynamic gain-scheduling controller. Using engine speed and oil pressure as the system parameters, the family of linear models was translated into an LPV system. An observer-based gain-scheduling controller for the LPV system is then designed based on the LMI technique. A discussion on weighting function selection for mixed $\mathscr{H}_2/\mathscr{H}_\infty$ controller synthesis is presented, with an emphasis placed on examining various frequency responses of the system. Test bench results show the effectiveness of the proposed scheme.

1.3 Notation and Preliminaries

Standard notation is used throughout this book. Let $\mathbb{R}$, $\mathbb{Z}$, and $\mathbb{Z}_{\geq 0}$ denote, respectively, the set of real, integer, and non-negative integer numbers. The positive definiteness of a matrix A is denoted by $A \succ 0$. The maximum (respectively, minimum) of α is denoted by $\bar{\alpha}$ (respectively, $\underline{\alpha}$). The abbreviation LFT is used to denote a linear fractional transformation, which is described in Appendix A. The ℓ_2 space of square-summable sequences on the set of nonnegative integers $\mathbb{Z}_{\geq 0}$ is given by

$$\ell_2 := \left\{ x : \mathbb{Z}_{\geq 0} \rightarrow \mathbb{R}^n \,\Big|\, \sum_{k=0}^{\infty} x^T(k)x(k) < \infty \right\}.$$

For a signal x in the ℓ_2 space, its ℓ_2 norm is defined as

$$\|x\|_{\ell_2} := \left(\sum_{k=0}^{\infty} x^T(k)x(k) \right)^{1/2}.$$

Other notation will be explained in due course.

1.4 Organization

This book, as depicted in Fig. 1.1, is organized as follows: a review of existing LPV control synthesis techniques and the modeling required to utilize them is presented in Chap. 2. These techniques are extended in Chap. 3 such that hard constraints on system outputs can be obtained with the guaranteed ℓ_2 to ℓ_∞ gain control problem. In Chap. 4 a gain-scheduled air-to-fuel ratio controller for port-fuel-injection engines is developed using the wall-wetting parameters and engine speed as the time-varying parameters for the LPV control synthesis. Results of both a simulation study and an HIL simulation are presented. In Chap. 5, a family of LTI systems, representing

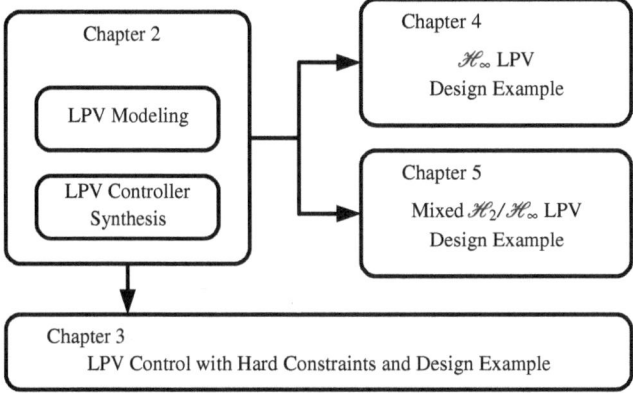

Fig. 1.1 Organization of the book

a variable valve timing actuator, obtained through closed-loop system identification [48, 49] are converted into an LPV model. LPV control synthesis is then applied to the LPV model to obtain a gain-scheduling controller for the variable valve timing actuator. The obtained controller is then validated on the variable valve timing actuator used for the system identification.

Chapter 2
Linear Parameter-Varying Modeling and Control Synthesis Methods

This chapter is split into the following two main parts: modeling of LPV systems and control synthesis methods for LPV systems.

2.1 Modeling of LPV Systems

Throughout this book, the control synthesis methods used rely on the existence of an LPV model with polytopic parameter dependence. Unfortunately, this is not the most intuitive form that an LPV model can take. Many physical systems have parameter variations that can be easily represented with LFTs. For this reason, we will demonstrate how to convert an LPV model with LFT parameter dependency into an LPV model with polytopic parameter dependence.

Consider the following open-loop, discrete-time LPV system with LFT parameter dependency:

$$
\begin{bmatrix} x(k+1) \\ l(k) \\ z(k) \\ y(k) \end{bmatrix} = \begin{bmatrix} A & B_p & B_w & B_u \\ C_l & D_{lp} & D_{lw} & D_{lu} \\ C_z & D_{zp} & D_{zw} & D_{zu} \\ C_y & D_{yp} & D_{yw} & D_{yu} \end{bmatrix} \begin{bmatrix} x(k) \\ p(k) \\ w(k) \\ u(k) \end{bmatrix}
\tag{2.1}
$$
$$
p(k) = \Theta(k)l(k)
$$

where $x(k)$ is the state at time k, $w(k)$ is the exogenous input, and $u(k)$ is the control input. The vectors $z(k)$ and $y(k)$ are the performance output and the measurement for control. Also, $p(k)$ and $l(k)$ are the pseudo-input and pseudo-output connected by $\Theta(k)$. The time-varying parameter $\Theta(k)$ follows the structure

$$
\Theta(k) \in \Theta = \left\{ \mathrm{diag}(\theta_1 I_{n_1}, \theta_2 I_{n_2}, \cdots, \theta_N I_{n_N}) \right\}.
\tag{2.2}
$$

A. P. White et al., *Linear Parameter-Varying Control for Engineering Applications*, SpringerBriefs in Control, Automation and Robotics, DOI: 10.1007/978-1-4471-5040-4_2, © The Author(s) 2013

Fig. 2.1 Diagram of the upper LFT of the state space matrices

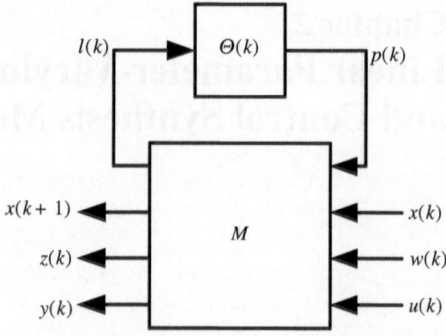

To emphasize the fact that there exists an LFT with respect to the time-varying parameter matrix $\Theta(k)$, the state-space matrices can be rearranged into the following upper LFT (Fig. 2.1):

$$
\begin{bmatrix} l(k) \\ x(k+1) \\ z(k) \\ y(k) \end{bmatrix} = \underbrace{\begin{bmatrix} D_{lp} & C_l & D_{lw} & D_{lu} \\ B_p & A & B_w & B_u \\ D_{zp} & C_z & D_{zw} & D_{zu} \\ D_{yp} & C_y & D_{yw} & D_{yu} \end{bmatrix}}_{=:M} \begin{bmatrix} p(k) \\ x(k) \\ w(k) \\ u(k) \end{bmatrix}
\tag{2.3}
$$

$$
p(k) = \Theta(k)l(k).
$$

The time-varying matrix $\Theta(k)$ can be absorbed back into the state space matrices such that the state space matrices would be given by

$$
\begin{bmatrix} x(k+1) \\ z(k) \\ y(k) \end{bmatrix} = \underbrace{\begin{bmatrix} A(\Theta(k)) & B_w(\Theta(k)) & B_u(\Theta(k)) \\ C_z(\Theta(k)) & D_{zw}(\Theta(k)) & D_{zu}(\Theta(k)) \\ C_y(\Theta(k)) & D_{yw}(\Theta(k)) & D_{yu}(\Theta(k)) \end{bmatrix}}_{=:H(\Theta)} \begin{bmatrix} x(k) \\ w(k) \\ u(k) \end{bmatrix}
\tag{2.4}
$$

where

$$
H(\Theta) := \mathscr{F}_u(M, \Theta)
$$

$$
= \begin{bmatrix} A & B_w & B_u \\ C_z & D_{zw} & D_{zu} \\ C_y & D_{yw} & D_{yu} \end{bmatrix} + \begin{bmatrix} B_p \\ D_{zp} \\ D_{yp} \end{bmatrix} \Theta(k) \left(I - D_{lp}\Theta(k) \right)^{-1} \begin{bmatrix} C_l & D_{lw} & D_{lu} \end{bmatrix}.
$$

$$
\tag{2.5}
$$

It is clear from (2.5) that when the matrix D_{lp} is non-zero, then the system matrices are not affine functions, i.e., a linear combination of the time-varying parameters plus a constant translation. Since, as previously mentioned, all control synthesis methods covered in this book rely on an LPV model with a polytopic parameter dependence, the system matrices must be affine functions of the time-varying parameters. If the matrix D_{lp} is non-zero, then some approximation must be made. If the parameter variation is "small", then a first-order Taylor series approximation can be performed.

2.1.1 First-Order Taylor Series Approximation of LPV Systems

Using the first-order Taylor series expansion at $\Theta = \bar{\Theta}$, the LPV system can be approximated as

$$\hat{H}(\Theta(k)) = H(\bar{\Theta}) + \sum_{i=1}^{N} \left[\nabla H(\bar{\Theta})\right]_i (\theta_i(k) - \bar{\theta}_i) \tag{2.6}$$

where $\theta_i(k)$, for $i = 1, \ldots, N$ are the individual parameters in $\Theta(k)$, and $\left[\nabla H(\bar{\Theta})\right]_i$ is the partial derivative of the LFT system $H(\Theta)$ with respect to θ_i solved at $\bar{\Theta}$. The i-th partial derivative of the upper LFT system $H(\Theta)$ is computed by [39]

$$[\nabla H(\Theta)]_i = M_{21}[I - \Theta M_{11}]^{-1} E_i [I - M_{11}\Theta]^{-1} M_{12}, \tag{2.7}$$

where

$$M_{11} = D_{lp}, \quad M_{12} = \begin{bmatrix} C_l & D_{lw} & D_{lu} \end{bmatrix}, \quad M_{21} = \begin{bmatrix} B_p \\ D_{zp} \\ D_{yp} \end{bmatrix}, \tag{2.8}$$

and the matrices E_i are defined such that

$$\Theta(k) = \sum_{i=1}^{N} \theta_i(k) E_i. \tag{2.9}$$

After performing this first-order Taylor series approximation, then the approximated system $\hat{H}(\Theta(k))$ will have affine parameter dependence with respect to $\Theta(k)$. As shown in the next section, a polytopic LPV model can be obtained from an LPV system with affine parameter dependence.

2.1.2 Polytopic Linear Time-Varying System with Barycentric Coordinates

The LPV system with affine parameter dependence can be represented by the following polytopic linear time-varying (LTV) system

$$\begin{bmatrix} x(k+1) \\ z(k) \\ y(k) \end{bmatrix} = \underbrace{\begin{bmatrix} A(\lambda(k)) & B_w(\lambda(k)) & B_u(\lambda(k)) \\ C_z(\lambda(k)) & D_{zw}(\lambda(k)) & D_{zu}(\lambda(k)) \\ C_y(\lambda(k)) & D_{yw}(\lambda(k)) & D_{yu}(\lambda(k)) \end{bmatrix}}_{=:H(\lambda)} \begin{bmatrix} x(k) \\ w(k) \\ u(k) \end{bmatrix} \tag{2.10}$$

where the system matrices $A(\lambda(k))$, $B_w(\lambda(k))$, $B_u(\lambda(k))$, $C_z(\lambda(k))$, $C_y(\lambda(k))$, $D_{zw}(\lambda(k))$, $D_{zu}(\lambda(k))$, $D_{yw}(\lambda(k))$, and $D_{yu}(\lambda(k))$ belong to the polytope

$$\begin{aligned} \mathfrak{D} = \Big\{ & (A, B_w, B_u, C_z, C_y, D_{zw}, D_{zu}, D_{yw}, D_{yu})(\lambda_k) : \\ & (A, B_w, B_u, C_z, C_y, D_{zw}, D_{zu}, D_{yw}, D_{yu})(\lambda_k) \tag{2.11} \\ & = \sum_{i=1}^{N} \lambda_i(k)(A, B_w, B_u, C_z, C_y, D_{zw}, D_{zu}, D_{yw}, D_{yu})_i, \lambda_k \in \Lambda_N \Big\}, \end{aligned}$$

with $(A, B_w, B_u, C_z, C_y, D_{zw}, D_{zu}, D_{yw}, D_{yu})_i$ the vertices of the polytope and $\lambda_k = \lambda(k) \in \mathbb{R}^N$ the vector of time-varying barycentric coordinates lying in the unit simplex

$$\Lambda_N = \Big\{ \zeta \in \mathbb{R}^N : \sum_{i=1}^{N} \zeta_i = 1, \zeta_i \geq 0, i = 1, \cdots, N \Big\}. \tag{2.12}$$

The vertices of the polytope $\mathfrak{D}$ are obtained by solving the system matrices of $\hat{H}(\Theta)$ at each of the vertices $\mathcal{V}_i$ for $i = 1, \ldots, N$ (See examples in Fig. 2.2). Then each of the state space matrices in $H(\lambda_k)$ are computed as the convex combination of the vertice systems of this polytope, such that, for example, the state matrix would be computed by

$$A(\lambda_k) = \sum_{i=1}^{N} \lambda_i(k) A_i. \tag{2.13}$$

Each of the other matrices in $H(\lambda_k)$ are computed in the same way. The convex combination coefficients $\{\lambda_i(\Theta)\}$ for a given Θ and set of vertices $\{\mathcal{V}_i\}$ are also known as the barycentric coordinates. The barycentric coordinate function is defined in [60] as

$$\lambda_i(\Theta) = \frac{\Upsilon_i(\Theta)}{\sum_i \Upsilon_i(\Theta)}, \tag{2.14}$$

where $\Upsilon_i(\Theta)$ is the weight function of vertex i for a point Θ inside of the convex polytope. The weight function is

$$\Upsilon_i(\Theta) = \frac{\text{vol}(\mathcal{V}_i)}{\Pi_{j \in \text{ind}(\mathcal{V}_i)}(n_j \cdot (\mathcal{V}_i - \Theta))}, \tag{2.15}$$

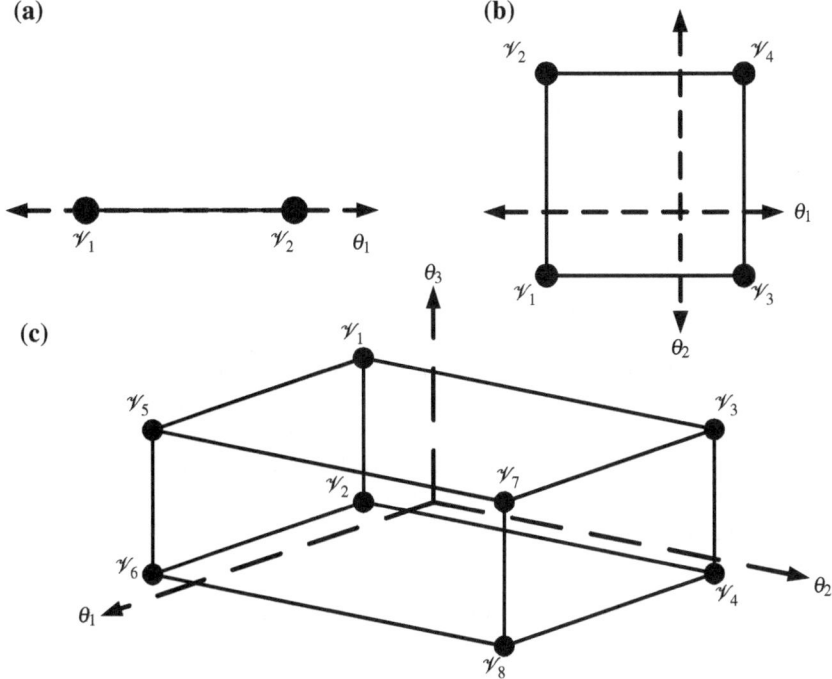

Fig. 2.2 Examples of possible parameter space polytopes

where $\mathrm{vol}(\mathscr{V}_i)$ is the volume of the parallelepiped span by the normals to the facets incident on vertex i, i.e., $\mathscr{V}_i$, $\{n_j\}$ is the collection of normal vectors to the facets incident on vertex i, and $\mathrm{ind}(\mathscr{V}_i)$ denotes the set of indices j such that the facet normal to n_j contains the vertex $\mathscr{V}_i$. The volume of a parallelepiped can be found as

$$\mathrm{vol}(\mathscr{V}_i) = |\det(n_{\mathrm{ind}})| . \tag{2.16}$$

where n_{ind} is a matrix whose rows are the vectors n_j where $j \in \mathrm{ind}(\mathscr{V}_i)$.

Since the polytopic LTV system has been defined, we will now focus our attention in the next section on the performance specifications for the polytopic LTV system.

2.2 Performance of Discrete-Time Polytopic LPV Systems

Consider the $\mathscr{H}_2$ or $\mathscr{H}_\infty$ weighted closed-loop discrete-time LPV system

$$H := \begin{cases} x(k+1) = \mathscr{A}(\lambda_k)x(k) + \mathscr{B}_w(\lambda_k)w(k), & x(0) = 0 \\ z(k) = \mathscr{C}_z(\lambda_k)x(k) + \mathscr{D}_w(\lambda_k)w(k) \end{cases} \tag{2.17}$$

where $x(k) \in \mathbb{R}^n$ is the state, $w(k) \in \mathbb{R}^r$ is the exogenous input, and $z(k) \in \mathbb{R}^p$ is the performance output. The system matrices $\mathscr{A}(\lambda_k)$, $\mathscr{B}_w(\lambda_k)$, $\mathscr{C}_z(\lambda_k)$, and $\mathscr{D}_w(\lambda_k)$ belong to a polytope similar to $\mathfrak{D}$ in (2.11).

The $\mathscr{H}_\infty$ performance of the system (2.17) from $w(k)$ to $z(k)$ is defined by the quantity

$$\|H\|_\infty = \sup_{\|w(k)\|_2 \neq 0} \frac{\|z(k)\|_2}{\|w(k)\|_2} \tag{2.18}$$

with $w(k) \in \ell_2$ and $z(k) \in \ell_2$. In robust control, the $\mathscr{H}_\infty$ norm has proved to be extremely useful and has various interpretations [54]. For example, in the frequency domain, the $\mathscr{H}_\infty$ norm is the peak value of the transfer function magnitude, for a single-input single-output (SISO) system. On the other hand, in the time domain the $\mathscr{H}_\infty$ norm can be thought of as the worst-case gain for sinusoidal inputs at any frequency. The $\mathscr{H}_\infty$ norm has been extremely useful in robust and LPV control because it is convenient for representing unstructured model uncertainties, and can therefore be useful in gaging the robustness of a system. By using the bounded real lemma, an upper bound for the $\mathscr{H}_\infty$ performance is characterized by the following lemma [13].

Lemma 2.1 (*$\mathscr{H}_\infty$ Performance*) *Consider the system H given by (2.17). If there exist bounded matrices $G(\lambda_k)$ and $P(\lambda_k) = P^T(\lambda_k) > 0$ for all $\lambda_k \in \Lambda_N$ such that*

$$\begin{bmatrix} P(\lambda_{k+1}) & \mathscr{A}(\lambda_k)G(\lambda_k) & \mathscr{B}_w(\lambda_k) & 0 \\ G^T(\lambda_k)\mathscr{A}^T(\lambda_k) & G(\lambda_k) + G^T(\lambda_k) - P(\lambda_k) & 0 & G^T(\lambda_k)\mathscr{C}_z^T(\lambda_k) \\ \mathscr{B}_w^T(\lambda_k) & 0 & \eta I & \mathscr{D}_w^T(\lambda_k) \\ 0 & \mathscr{C}_z(\lambda_k)G(\lambda_k) & \mathscr{D}_w(\lambda_k) & \eta I \end{bmatrix} > 0 \tag{2.19}$$

then the system H is exponentially stable and

$$\|H\|_\infty \leq \inf_{P(\lambda_k),G(\lambda_k),\eta} \eta.$$

This lemma is an extension of a standard result provided by [18, 19].

The $\mathscr{H}_2$ norm has two main interpretations: deterministic and stochastic, depending on what type of input signal is considered. For the deterministic interpretation, the input signal has bounded energy (ℓ_2 norm) and the $\mathscr{H}_2$ norm is the peak magnitude (or ℓ_∞ norm) of the performance output divided by the energy of the input. For the stochastic interpretation, the input signal is assumed to be white noise with unit intensity and the $\mathscr{H}_2$ norm is then the energy of the performance output (ℓ_2 norm) [54]. For discrete-time LTI systems, there are three main definitions that are usually used to define the $\mathscr{H}_2$ norm [12, 13, 55]. They are as follows:

1. If $H(q)$ represents the transfer function matrix of a system $H(q)$, then its $\mathscr{H}_2$ norm can be defined as

$$\|H(q)\|_2^2 = \frac{1}{2\pi} \int\limits_0^{2\pi} \text{trace}\left\{H^T(e^{j\omega})H(e^{j\omega})\right\} d\omega. \tag{2.20}$$

2. If $\{e_1, \ldots, e_r\}$ is a basis for the input space and $z_i(k)$ is the output of the system H when an impulse $\delta(k)e_i$ is applied, then its $\mathcal{H}_2$ norm can be defined as

$$\|H\|_2^2 = \sum_{i=1}^r \|z_i\|_2^2. \tag{2.21}$$

3. If $z(k)$ is the output of an LTI system when a zero-mean white noise Gaussian process $w(k)$ with identity covariance matrix is applied, then its $\mathcal{H}_2$ norm can be defined as

$$\|H\|_2^2 = \lim_{m \to \infty} \sup \mathscr{E}\left\{\frac{1}{m}\sum_{k=0}^m z^T(k)z(k)\right\} \tag{2.22}$$

where $\mathscr{E}$ denotes the expectation operator and the positive integer m denotes the time horizon.

Since the idea of a transfer function is not well defined for time-varying systems, the first definition has not been extended to LTV systems. The second and third definitions can be extended to LTV systems. However, since the computation of the norm with the second definition can depend on the selection of the basis for the input space, the third definition has received more attention [8, 12, 13, 27, 55]. An upper bound for the $\mathcal{H}_2$ performance given by the third definition is characterized by the following lemma [13].

Lemma 2.2 ($\mathcal{H}_2$ **Performance**) *Consider the system H given by (2.17). If there exists bounded matrices $G(\lambda_k)$, $P(\lambda_k) = P^T(\lambda_k) > 0$, and $W(\lambda_k) = W^T(\lambda_k)$ for all $\lambda_k \in \Lambda_N$ such that*

$$\begin{bmatrix} P(\lambda_{k+1}) & \mathscr{A}(\lambda_k)G(\lambda_k) & \mathscr{B}_w(\lambda_k) \\ G^T(\lambda_k)\mathscr{A}^T(\lambda_k) & G(\lambda_k) + G^T(\lambda_k) - P(\lambda_k) & 0 \\ \mathscr{B}_w^T(\lambda_k) & 0 & I \end{bmatrix} > 0 \tag{2.23}$$

and

$$\begin{bmatrix} W(\lambda_k) - \mathscr{D}_w(\lambda_k)\mathscr{D}_w^T(\lambda_k) & \mathscr{C}_z(\lambda_k)G(\lambda_k) \\ G^T(\lambda_k)\mathscr{C}_z^T(\lambda_k) & G(\lambda_k) + G^T(\lambda_k) - P(\lambda_k) \end{bmatrix} > 0 \tag{2.24}$$

then the system H is exponentially stable and its $\mathcal{H}_2$ performance is bounded by v given by

$$v^2 = \inf_{P(\lambda_k),G(\lambda_k),W(\lambda_k)} \sup_{\lambda_k \in \Lambda_N} \text{trace}\{W(\lambda_k)\}$$

such that $\|H\|_2 \leq v$.

The proof for this lemma can be found in [13].

Note that the parameter-dependent LMI conditions in Lemmas 2.1 and 2.2 must be evaluated for all λ_k in the unit simplex Λ_N. This leads to an infinite dimensional problem. By imposing an affine parameter-dependent structure on the Lyapunov matrix $P(\lambda_k)$, such that

$$P(\lambda_k) = \sum_{i=1}^{N} \lambda_i(k) P_i, \quad i = 1, \ldots, N, \tag{2.25}$$

a finite set of LMIs in terms of the vertices of the polytope $\mathfrak{D}$ can be obtained.

To reduce conservatism, the parameter variation rate

$$\Delta\lambda_i(k) = \lambda_i(k+1) - \lambda_i(k), \quad i = 1, \ldots, N \tag{2.26}$$

is assumed to be limited. Two limits have been considered in the literature. The first rate limit considered in the literature [11, 12, 40] is given by

$$-b\lambda_i(k) \le \Delta\lambda_i(k) \le b\left(1 - \lambda_i(k)\right), \quad i = 1, \ldots, N, \tag{2.27}$$

with $b \in [0, 1]$. With this parameter variation rate bound and the affine parameter-dependent structure in (2.25), the $\mathcal{H}_\infty$ performance criteria in Lemma 2.1 can be transformed into a finite number of LMIs, as shown in the next Lemma [11].

Lemma 2.3 (*Finite $\mathcal{H}_\infty$ Performance with rate limit* (2.27)) *The system H* (2.17) *has an $\mathcal{H}_\infty$ performance bounded by η if there exist matrices $G_i \in \mathbb{R}^{n \times n}$ and symmetric matrices $P_i \in \mathbb{R}^{n \times n}$ such that*

$$\begin{bmatrix} (1-b)P_i + bP_\ell & \star & \star & \star \\ G_i^T \mathcal{A}_i^T & G_i + G_i^T - P_i & \star & \star \\ \mathcal{B}_{w,i}^T & 0 & \eta I & \star \\ 0 & \mathcal{C}_{z,i} G_i & \mathcal{D}_{w,i} & \eta I \end{bmatrix} > 0 \tag{2.28}$$

holds for $i = 1, \ldots, N$ and $\ell = 1, \ldots, N$ and

$$\begin{bmatrix} (1-b)P_i + (1-b)P_j + 2bP_\ell & \star & \star & \star \\ G_j^T \mathcal{A}_i^T + G_i^T \mathcal{A}_j^T & G_i + G_i^T + G_j + G_j^T - P_i - P_j & \star & \star \\ \mathcal{B}_{w,i}^T + \mathcal{B}_{w,j}^T & 0 & 2\eta I & \star \\ 0 & \mathcal{C}_{z,i} G_j + \mathcal{C}_{z,j} G_i & \mathcal{D}_{w,i} + \mathcal{D}_{w,j} & 2\eta I \end{bmatrix}$$
$$> 0 \tag{2.29}$$

holds for $\ell = 1, \ldots, N$, $i = 1, \ldots, N-1$, and $j = i+1, \ldots, N$.

A proof for this lemma can be found in [11].

Likewise, the $\mathcal{H}_2$ performance criteria in Lemma 2.2 can also be transformed into a finite number of LMIs as shown in [12] for the case when $\mathcal{D}_w = 0$.

Lemma 2.4 (**Finite $\mathcal{H}_2$ Performance with rate limit** (2.27)) *Consider the system H (2.17), with $\mathcal{D}_w = 0$. If there exist matrices $G_i \in \mathbb{R}^{n \times n}$ and symmetric matrices $P_i \in \mathbb{R}^{n \times n}$ and $W_i \in \mathbb{R}^{p \times p}$ such that*

$$\begin{bmatrix} (1-b)P_i + bP_\ell & \star & \star \\ G_i^T \mathscr{A}_i^T & G_i + G_i^T - P_i & \star \\ \mathscr{B}_{w,i}^T & 0 & I \end{bmatrix} > 0, \qquad (2.30)$$

for $i = 1, \ldots, N$ and $\ell = 1, \ldots, N$,

$$\begin{bmatrix} (1-b)\left(P_i + P_j\right) + 2bP_\ell & \star & \star \\ G_j^T \mathscr{A}_i^T + G_i^T \mathscr{A}_j^T & G_i + G_i^T + G_j + G_j^T - P_i - P_j & \star \\ \mathscr{B}_{w,i}^T + \mathscr{B}_{w,j}^T & 0 & 2I \end{bmatrix} > 0 \quad (2.31)$$

for $\ell = 1, \ldots, N$, $i = 1, \ldots, N-1$, and $j = i+1, \ldots, N$,

$$\begin{bmatrix} W_i & \star \\ G_i^T \mathscr{C}_{z,i}^T & G_i + G_i^T - P_i \end{bmatrix} > 0 \qquad (2.32)$$

for $i = 1, \ldots, N$, and

$$\begin{bmatrix} W_i + W_j & \star \\ G_j^T \mathscr{C}_{z,i}^T + G_i^T \mathscr{C}_{z,j}^T & G_i + G_i^T + G_j + G_j^T - P_i - P_j \end{bmatrix} > 0 \qquad (2.33)$$

for $i = 1, \ldots, N-1$ and $j = i+1, \ldots, N$, then the system H, with $\mathcal{D}_w = 0$, is exponentially stable and has an $\mathcal{H}_2$ performance bounded by v given by

$$v^2 = \min_{G_i, P_i, W_i} \max_i \text{trace}\{W_i\}. \qquad (2.34)$$

A proof for this lemma can be found in [12].

While the rate limit (2.27) can be useful, it may or may not be very realistic. To see this, one may consider the example parameter variation with $N = 2$ and $b = 0.5$ as displayed in Fig. 2.3. In this example, the time-varying parameter starts at one extreme and moves the other extreme as quickly as the parameter variation rate limit (2.27) allows. It is clear that the maximum parameter variation rate is dependent on the current value of the parameters with the rate limit given by (2.27).

A more realistic parameter variation limit that is not dependent on the current value of the time-varying parameter is considered in [13, 42]. This limit is given by

$$-b \le \Delta\lambda_i(k) \le b, \quad i = 1, \ldots, N, \qquad (2.35)$$

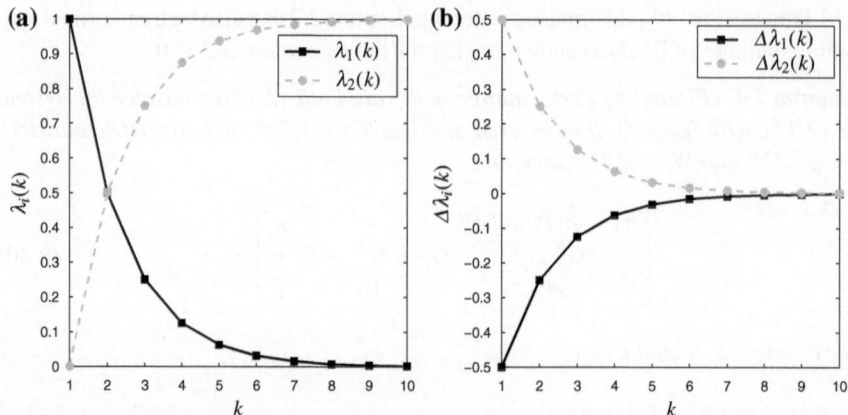

Fig. 2.3 Fastest possible parameter transition between the extreme conditions with $N = 2$ and $b = 0.5$ when the rate limit given by (2.27) is in effect

with $b \in [0, 1]$. When using this parameter variation rate, the uncertainty domain, where the vector $(\lambda(k), \Delta\lambda(k))^T \in \mathbb{R}^{2N}$ takes values, may be modeled by the compact set

$$\Gamma_b = \left\{ \delta \in \mathbb{R}^{2N} : \delta \in \text{co}\{g^1, \ldots, g^M\}, g^j = \begin{pmatrix} f^j \\ h^j \end{pmatrix}, \quad f^j \in \mathbb{R}^N, \quad h^j \in \mathbb{R}^N, \right.$$

$$\left. \sum_{i=1}^{N} f_i^j = 1 \text{ with } f_i^j \geq 0, i = 1, \ldots, N, \sum_{i=1}^{N} h_i^j = 0, \quad j = 1, \ldots, M \right\} \tag{2.36}$$

defined as the convex combination of the vectors g^j, for $j = 1, \ldots, M$, given a priori. This definition of Γ_b ensures that $\lambda(k) \in \Lambda_N$ and that

$$\sum_{i=1}^{N} \Delta\lambda_i(k) = 0 \tag{2.37}$$

holds for all $k \geq 0$. For a given bound b, the columns of Γ_b can be generated as the columns of a matrix V as follows [13] (in MATLAB code)

```
V = zeros(2*N,N^2+(N-1)^2+(N-1));
for i = 1:1:N
    V(i,(i-1)*N+1) = 1;
    ind = 1;
    for j = 1:1:N
        if j ISNOT i
            V(i,(i-1)*N+ind+1) = 1;
```

```
V(N+i,(i-1)*N+ind+1)  = -b;
V(N+j,(i-1)*N+ind+1)  = b;

V(i,N^2+(i-1)*(N-1)+ind)    = b;
V(j,N^2+(i-1)*(N-1)+ind)    = 1-b;
V(N+i,N^2+(i-1)*(N-1)+ind)  = -b;
V(N+j,N^2+(i-1)*(N-1)+ind)  = b;

ind = ind + 1;
        end
    end
end
f = V(1:N,:);
h = V(N+1:2*N,:);
```

With the uncertainty set Γ_b, each $\lambda_i(k)$ and $\Delta\lambda_i(k)$ for $i = 1, 2, \ldots, N$ are given by

$$\lambda_i(k) = \sum_{j=1}^{M} f_i^j \gamma_j(k) \quad \text{and} \quad \Delta\lambda_i(k) = \sum_{j=1}^{M} h_i^j \gamma_j(k) \tag{2.38}$$

such that the affine representation of $P(\lambda_k)$ is given by

$$P(\lambda_k) = \sum_{i=1}^{N} \lambda_i(k) P_i = \sum_{i=1}^{N} \left(\sum_{j=1}^{M} f_i^j \gamma_j(k) \right) P_i$$

$$= \sum_{j=1}^{M} \gamma_j(k) \left(\sum_{i=1}^{N} f_i^j P_i \right) = \sum_{j=1}^{M} \gamma_j(k) \tilde{P}_j = \tilde{P}(\gamma(k)) \tag{2.39}$$

with $\tilde{P}_j = \sum_{i=1}^{N} f_i^j P_i$ as shown in [13]. Using the same structure for λ_k, the system matrices in H (2.17) are also converted to the new representation in terms of $\gamma(k) \in \Lambda_M$, such that

$$\mathscr{A}(\lambda_k) = \tilde{\mathscr{A}}(\gamma(k)) = \sum_{j=1}^{M} \gamma_j(k) \tilde{\mathscr{A}}_j \tag{2.40}$$

with $\tilde{\mathscr{A}}_j = \sum_{i=1}^{N} f_i^j \mathscr{A}_i$. All other matrices in H are converted in the same way. It is also shown in [13], that by combining (2.38) with the fact that $\lambda_{k+1} = \lambda_k + \Delta\lambda_k$,

$$P(\lambda_{k+1}) = \sum_{i=1}^{N} (\lambda_i(k) + \Delta\lambda_i(k)) P_i = \sum_{i=1}^{N} \left(\sum_{j=1}^{M} \left(f_i^j + h_i^j \right) \gamma_j(k) \right) P_i$$

$$= \sum_{j=1}^{M} \gamma_j(k) \left(\sum_{i=1}^{N} \left(f_i^j + h_i^j \right) P_i \right) = \sum_{j=1}^{M} \gamma_j(k) \hat{P}_j = \hat{P}(\gamma(k)) \quad (2.41)$$

with $\hat{P}_j = \sum_{i=1}^{N} \left(f_i^j + h_i^j \right) P_i$. The authors of [13] note that due to these representations of $P(\lambda_k)$ and $P(\lambda_{k+1})$, the LMIs of Lemma 2.1 and Lemma 2.2 can be rewritten with a dependency on $\gamma(k)$. They also note that a convenient parameterization of the slack variable $G(\lambda_k)$ is given by

$$G(\lambda_k) = G(\gamma(k)) = \sum_{j=1}^{M} \gamma_j(k) G_j, \quad \gamma(k) \in \Lambda_M. \quad (2.42)$$

Using these parameterizations, the next two lemmas present a finite-dimensional set of LMIs that guarantee the LMI conditions of Lemmas 2.1 and 2.2 [13].

Lemma 2.5 (*Finite $\mathcal{H}_\infty$ Performance with rate limit* (2.35)) *Consider the system H given by* (2.17). *Assume that the vectors f^j and h^j of Γ_b are given. If there exist, for $j = 1, \ldots, M$, matrices $G_j \in \mathbb{R}^{n \times n}$ and, for $i = 1, \ldots, N$, symmetric positive-definite matrices $P_i \in \mathbb{R}^{n \times n}$ such that*

$$\begin{bmatrix} \sum_{i=1}^{N} \left(f_i^j + h_i^j \right) P_i & \star & & \star & \star \\ G_j^T \tilde{\mathscr{A}}_j^T & G_j + G_j^T - \sum_{i=1}^{N} f_i^j P_i & \star & \star \\ \tilde{\mathscr{B}}_{w,j}^T & 0 & \eta I & \star \\ 0 & \tilde{\mathscr{C}}_{z,j} G_j & \tilde{\mathscr{D}}_{w,j} & \eta I \end{bmatrix} > 0 \quad (2.43)$$

for $j = 1, \ldots, M$ and

$$\begin{bmatrix} \sum_{i=1}^{N} \left(f_i^j + f_i^\ell + h_i^j + h_i^\ell \right) P_i & \star & & \star & \star \\ G_j^T \tilde{\mathscr{A}}_\ell^T + G_\ell^T \tilde{\mathscr{A}}_j^T & \Theta_{22,j\ell} & \star & \star \\ \tilde{\mathscr{B}}_{w,j}^T + \tilde{\mathscr{B}}_{w,\ell}^T & 0 & 2\eta I & \star \\ 0 & \tilde{\mathscr{C}}_{z,j} G_\ell + \tilde{\mathscr{C}}_{z,\ell} G_j & \tilde{\mathscr{D}}_{w,j} + \tilde{\mathscr{D}}_{w,\ell} & 2\eta I \end{bmatrix} > 0 \quad (2.44)$$

with

$$\Theta_{22,j\ell} = G_j + G_j^T + G_\ell + G_\ell^T - \sum_{i=1}^{N} \left(f_i^j + f_i^\ell \right) P_i$$

for $j = 1, \ldots, M - 1$ and $\ell = j + 1, \ldots, M$, then the system H is exponentially stable and

$$\|H\|_\infty \leq \min_{P_i, G_j, \eta} \eta.$$

The proof for this lemma can be found in [13].

Lemma 2.6 (*Finite $\mathcal{H}_2$ Performance with rate limit* (2.35)) *Consider the system H given by (2.17). Assume that the vectors f^j and h^j of Γ_b are given. If there exist, for $j = 1, \ldots, M$, matrices $G_j \in \mathbb{R}^{n \times n}$ and, for $i = 1, \ldots, N$, symmetric positive-definite matrices $P_i \in \mathbb{R}^{n \times n}$ and $W_i \in \mathbb{R}^{p \times p}$ such that*

$$\begin{bmatrix} \sum_{i=1}^{N} \left(f_i^j + h_i^j \right) P_i & \star & \star \\ G_j^T \tilde{\mathscr{A}}_j^T & G_j + G_j^T - \sum_{i=1}^{N} f_i^j P_i & \star \\ \mathscr{B}_{w,j}^T & 0 & I \end{bmatrix} > 0 \qquad (2.45)$$

for $j = 1, \ldots, M$,

$$\begin{bmatrix} \sum_{i=1}^{N} \left(f_i^j + f_i^\ell + h_i^j + h_i^\ell \right) P_i & \star & \star \\ G_j^T \tilde{\mathscr{A}}_\ell^T + G_\ell^T \tilde{\mathscr{A}}_j^T & G_j + G_j^T + G_\ell + G_\ell^T - \sum_{i=1}^{N} \left(f_i^j + f_i^\ell \right) P_i & \star \\ \mathscr{B}_{w,j}^T + \mathscr{B}_{w,\ell}^T & 0 & 2I \end{bmatrix} \\ > 0 \qquad (2.46)$$

for $j = 1, \ldots, M - 1$, and $\ell = j + 1, \ldots, M$,

$$\begin{bmatrix} \sum_{i=1}^{N} f_i^j W_i - \tilde{\mathscr{D}}_{w,j} \tilde{\mathscr{D}}_{w,j}^T & \star \\ G_j^T \tilde{\mathscr{C}}_{z,j}^T & G_j + G_j^T - \sum_{i=1}^{N} f_i^j P_i \end{bmatrix} > 0 \qquad (2.47)$$

for $j = 1, \ldots, M$,

$$\begin{bmatrix} \sum_{i=1}^{N} \left(f_i^j + f_i^\ell \right) W_i - \tilde{\mathscr{D}}_{w,j} \tilde{\mathscr{D}}_{w,\ell}^T + \tilde{\mathscr{D}}_{w,\ell} \tilde{\mathscr{D}}_{w,j}^T & \star \\ G_j^T \tilde{\mathscr{C}}_{z,\ell}^T + G_\ell^T \tilde{\mathscr{C}}_{z,j}^T & G_j + G_j^T + G_\ell + G_\ell^T - \sum_{i=1}^{N} \left(f_i^j + f_i^\ell \right) P_i \end{bmatrix} \\ > 0 \qquad (2.48)$$

for $j = 1, \ldots, M - 1$, and $\ell = j + 1, \ldots, M$, then the system H is exponentially stable and its $\mathcal{H}_2$ performance is bound by v given by

$$v^2 = \min_{P_i, G_j, W_i} \max_i \text{trace} \{W_i\}.$$

The proof for this lemma can be found in [13].

2.3 Control Synthesis Methods for LPV Systems

In this section, the gain-scheduled static output feedback controller synthesis results from [11–13] are reviewed.

Consider the following $\mathcal{H}_\infty$ and $\mathcal{H}_2$ weighted, discrete-time polytopic time-varying systems H^∞ and H^2:

$$H^\infty := \begin{cases} x(k+1) = A(\lambda_k)x(k) + B_{\infty w}(\lambda_k)w_\infty(k) + B_u(\lambda_k)u(k) \\ z_\infty(k) = C_{\infty z}(\lambda_k)x(k) + D_{\infty w}(\lambda_k)w_\infty(k) + D_{\infty u}(\lambda_k)u(k) \\ y(k) = C_y x(k), \quad C_y = [I_q,\ 0] \end{cases} \quad (2.49)$$

$$H^2 := \begin{cases} x(k+1) = A(\lambda_k)x(k) + B_{2w}(\lambda_k)w_2(k) + B_u(\lambda_k)u(k) \\ z_2(k) = C_{2z}(\lambda_k)x(k) + D_{2w}(\lambda_k)w_2(k) + D_{2u}(\lambda_k)u(k) \\ y(k) = C_y x(k), \quad C_y = [I_q,\ 0] \end{cases} \quad (2.50)$$

where $x(k) \in \mathbb{R}^n$ is the state, $w_\infty(k) \in \mathbb{R}^{r_\infty}$ and $w_2(k) \in \mathbb{R}^{r_2}$ are the $\mathcal{H}_\infty$ and $\mathcal{H}_2$ exogenous inputs, $z_\infty(k) \in \mathbb{R}^{p_\infty}$ and $z_2(k) \in \mathbb{R}^{p_2}$ are the $\mathcal{H}_\infty$ and $\mathcal{H}_2$ performance outputs, and $y \in \mathbb{R}^q$ is the measurement for control. The system matrices of H^∞ and H^2 belong to a polytope similar to the one given in (2.11).

2.3.1 $\mathcal{H}_\infty$ Control Synthesis

In [11], a finite set of LMIs is presented which can be used to synthesize a stabilizing, static output feedback LPV controller for the system H^∞ with a guaranteed $\mathcal{H}_\infty$ performance bound. The rate of variation of the parameters (2.26) is assumed to be limited by (2.27).

Extending the analysis result presented in Lemma 2.3, the authors of [11] characterize a finite set of LMI conditions for the synthesis of a gain scheduled $\mathcal{H}_\infty$ static output feedback controller for the system (2.49).

Lemma 2.7 *Consider the system H^∞ given by (2.49). If there exist matrices $G_{i,1} \in \mathbb{R}^{q \times q}$, $G_{i,2} \in \mathbb{R}^{n-q,q}$, $G_{i,3} \in \mathbb{R}^{n-q \times n-q}$, $Z_i \in \mathbb{R}^{m \times q}$, and symmetric matrices $P_i \in \mathbb{R}^{n \times n}$ such that*

$$\begin{bmatrix} (1-b)P_i + bP_\ell & \star & \star & \star \\ G_i^T A_i^T + Z_i^T B_{u,i}^T & G_i + G_i^T - P_i & \star & \star \\ B_{\infty w,i}^T & 0 & \eta I & \star \\ 0 & C_{\infty z,i}G_i + D_{\infty u,i}Z_i & D_{\infty w,i} & \eta I \end{bmatrix} > 0 \quad (2.51)$$

hold for $i = 1, \ldots, N$ and $\ell = 1, \ldots, N$ and

$$\begin{bmatrix} (1-b)P_i + (1-b)P_j + 2bP_\ell & \star & \star & \star \\ \Theta_{21,ij} & G_i + G_i^T + G_j + G_j^T - P_i - P_j & \star & \star \\ B_{\infty w,i}^T + B_{\infty w,j}^T & 0 & 2\eta I & \star \\ 0 & \Theta_{42,ij} & D_{\infty w,i} + D_{\infty w,j} & 2\eta I \end{bmatrix} \\ > 0 \tag{2.52}$$

with

$$\Theta_{21,ij} = G_j^T A_i^T + G_i^T A_j^T + Z_j^T B_{u,i}^T + Z_i^T B_{u,j}^T$$
$$\Theta_{42,ij} = C_{\infty z,i} G_j + C_{\infty z,j} G_i + D_{\infty u,i} Z_j + D_{\infty u,j} Z_i$$

hold for $\ell = 1, \ldots, N$, $i = 1, \ldots, N-1$, and $j = i+1, \ldots, N$, with

$$G_i = \begin{bmatrix} G_{i,1} & 0 \\ G_{i,2} & G_{i,3} \end{bmatrix} \quad \text{and} \quad Z_i = \begin{bmatrix} Z_{i,1} & 0 \end{bmatrix},$$

then the parameter-dependent static output feedback gain

$$K(\lambda_k) = \hat{Z}(\lambda_k)\hat{G}(\lambda_k)^{-1}, \tag{2.53}$$

with

$$\hat{Z}(\lambda(k)) = \sum_{i=1}^{N} \lambda_i(k)Z_{i,1} \quad \text{and} \quad \hat{G}(\lambda(k)) = \sum_{i=1}^{N} \lambda_i(k)G_{i,1}$$

stabilizes the system (2.49) with a guaranteed $\mathscr{H}_\infty$ performance bounded by η for all $\lambda \in \Lambda_N$ and $\Delta\lambda$ that satisfies (2.27).

A proof for this lemma can be found in [11].

2.3.2 $\mathscr{H}_2$ Control Synthesis

In [12], a finite set of LMIs is presented which can be used to synthesize a stabilizing, static output feedback LPV controller for the system H^2 with a guaranteed $\mathscr{H}_2$ performance bound. The rate of variation of the parameters (2.26) is assumed to be limited by (2.27).

Extending the analysis result presented in Lemma 2.4, the authors of [12] characterize a finite set of LMI conditions for the synthesis of a gain scheduled $\mathscr{H}_2$ static output feedback controller for the system (2.50) with $D_{2w} = 0$.

Lemma 2.8 *Consider the system H^2 given by (2.49) with $D_{2w} = 0$. If there exist matrices $G_{i,1} \in \mathbb{R}^{q \times q}$, $G_{i,2} \in \mathbb{R}^{n-q,q}$, $G_{i,3} \in \mathbb{R}^{n-q \times n-q}$, $Z_i \in \mathbb{R}^{m \times q}$, and symmetric matrices $P_i \in \mathbb{R}^{n \times n}$ and $W_i \in \mathbb{R}^{p_2 \times p_2}$ such that*

$$
\begin{bmatrix}
(1-b)P_i + bP_\ell & \star & \star \\
G_i^T A_i^T + Z_i^T B_{u,i}^T & G_i + G_i^T - P_i & \star \\
B_{2w,i}^T & 0 & I
\end{bmatrix} > 0
\tag{2.54}
$$

for i = 1, ..., N and ℓ = 1, ..., N,

$$
\begin{bmatrix}
(1-b)P_i + (1-b)P_j + 2bP_\ell & \star & \star \\
G_j^T A_i^T + G_i^T A_j^T + Z_j^T B_{u,i}^T + Z_i^T B_{u,j}^T & G_i + G_i^T + G_j + G_j^T - P_i - P_j & \star \\
B_{2w,i}^T + B_{2w,j}^T & 0 & 2I
\end{bmatrix} > 0
\tag{2.55}
$$

for ℓ = 1, ..., N, i = 1, ..., N − 1, and j = i + 1, ..., N,

$$
\begin{bmatrix}
W_i & \star \\
G_i^T C_{2z,i}^T + Z_i^T D_{2u,i}^T & G_i + G_i^T - P_i
\end{bmatrix} > 0
\tag{2.56}
$$

for i = 1, ..., N, and

$$
\begin{bmatrix}
W_i + W_j & \star \\
G_j^T C_{2z,i}^T + G_i^T C_{2z,j}^T + Z_j^T D_{2u,i}^T + Z_i^T D_{2u,j}^T & G_i + G_i^T + G_j + G_j^T - P_i - P_j
\end{bmatrix} > 0
\tag{2.57}
$$

for i = 1, ..., N − 1 and j = i + 1, ..., N, with

$$
G_i = \begin{bmatrix} G_{i,1} & 0 \\ G_{i,2} & G_{i,3} \end{bmatrix} \quad and \quad Z_i = \begin{bmatrix} Z_{i,1} & 0 \end{bmatrix},
$$

then the parameter-dependent static output feedback gain

$$
K(\lambda_k) = \hat{Z}(\lambda_k)\hat{G}(\lambda_k)^{-1},
\tag{2.58}
$$

with

$$
\hat{Z}(\lambda(k)) = \sum_{i=1}^{N} \lambda_i(k)Z_{i,1} \quad and \quad \hat{G}(\lambda(k)) = \sum_{i=1}^{N} \lambda_i(k)G_{i,1}
$$

stabilizes the system (2.50) with a guaranteed $\mathcal{H}_2$ performance bounded by ν given by

$$
\nu^2 = \min_{G_i, Z_i, P_i, W_i} \max_i \text{trace}\{W_i\}.
\tag{2.59}
$$

for all λ ∈ Λ_N and Δλ that satisfies (2.27).

A proof for this lemma can be found in [12].

2.3.3 Mixed $\mathcal{H}_2/\mathcal{H}_\infty$ Control Synthesis

In [13], gain-scheduled static output feedback synthesis LMIs that stabilize the systems H^∞ and H^2 with $\mathcal{H}_\infty$ and $\mathcal{H}_2$ performance bounds are presented. The rate of variation of the parameters (2.26) is assumed to be limited by a priori known bound b, given by (2.35).

The authors of [13] extend the analysis results of Lemmas 2.5 and 2.6 to characterize a finite set of LMI conditions for the synthesis of a gain scheduled mixed $\mathcal{H}_2/\mathcal{H}_\infty$ static output feedback controller for the systems H^2 (2.50) and H^∞ (2.49).

Lemma 2.9 *Consider the systems H^∞ (2.49) and H^2 (2.50). Assume that the vectors f^j and h^j of Γ_b are given. Additionally, assume that a prescribed $\mathcal{H}_\infty$ performance bound η is given. If there exist, for $i = 1, \ldots, N$, matrices $G_{i,1} \in \mathbb{R}^{n\times n}$, $Z_{i,1} \in \mathbb{R}^{m\times q}$ and symmetric positive-definite matrices $P_{\infty,i} \in \mathbb{R}^{n\times n}$, $P_{2,i} \in \mathbb{R}^{n\times n}$, and $W_i \in \mathbb{R}^{p\times p}$, and, for $j = 1, \ldots, M$, matrices $G_{\infty j,2} \in \mathbb{R}^{(n-q)\times q}$, $G_{2j,2} \in \mathbb{R}^{(n-q)\times q}$, $G_{\infty j,3} \in \mathbb{R}^{(n-q)\times(n-q)}$, and $G_{2j,3} \in \mathbb{R}^{(n-q)\times(n-q)}$ such that*

$$
\begin{bmatrix}
\sum_{i=1}^{N}\left(f_i^j + h_i^j\right)P_{\infty,i} & \star & & \star & \star \\
G_{\infty,j}^T \tilde{A}_j^T + Z_j^T \tilde{B}_{u,j}^T & G_{\infty,j} + G_{\infty,j} - \sum_{i=1}^{N} f_i^j P_{\infty,i} & \star & \star \\
\tilde{B}_{\infty w,j}^T & 0 & \eta I & \star \\
0 & \tilde{C}_{\infty z,j}G_{\infty,j} + \tilde{D}_{\infty u,j}Z_j & \tilde{D}_{\infty w,j} & \eta I
\end{bmatrix} = \Theta_j > 0
$$

(2.60)

for $j = 1, \ldots, M$ and

$$
\begin{bmatrix}
\sum_{i=1}^{N}\left(f_i^j + f_i^\ell + h_i^j + h_i^\ell\right)P_{\infty,i} & \star & & \star & \star \\
\Theta_{21,j\ell} & \Theta_{22,j\ell} & \star & \star \\
\tilde{B}_{\infty w,j}^T + \tilde{B}_{\infty w,\ell}^T & 0 & 2\eta I & \star \\
0 & \Theta_{42,j\ell} & \tilde{D}_{\infty w,j} + \tilde{D}_{\infty w,\ell} & 2\eta I
\end{bmatrix} = \Theta_{j\ell} > 0
$$

(2.61)

with

$$
\Theta_{21,j\ell} = G_{\infty,j}^T \tilde{A}_\ell^T + G_{\infty,\ell}^T \tilde{A}_j^T + Z_j^T \tilde{B}_{u,\ell}^T + Z_\ell^T \tilde{B}_{u,j}^T
$$

$$
\Theta_{22,j\ell} = G_{\infty,j} + G_{\infty,j}^T + G_{\infty,\ell} + G_{\infty,\ell}^T - \sum_{i=1}^{N}\left(f_i^j + h_i^j\right)P_{\infty,i}
$$

$$
\Theta_{42,j\ell} = \tilde{C}_{\infty z,j}G_{\infty,\ell} + \tilde{C}_{\infty z,\ell}G_{\infty,j} + \tilde{D}_{u,j}Z_\ell + \tilde{D}_{u,\ell}Z_j
$$

for $j = 1, \ldots, M-1$ and $\ell = j+1, \ldots, M$, and

$$
\begin{bmatrix}
\sum_{i=1}^{N}\left(f_i^j + h_i^j\right)P_{2,i} & \star & \star \\
G_{2,j}^T \tilde{A}_j^T + Z_j^T \tilde{B}_{u,j}^T & G_{2,j} + G_{2,j}^T - \sum_{i=1}^{N} f_i^j P_{2,i} & \star \\
\tilde{B}_{w2,j}^T & 0 & I
\end{bmatrix} = \Phi_j > 0 \quad (2.62)
$$

for $j = 1, \ldots, M$, and

$$\begin{bmatrix} \sum_{i=1}^{N} \left(f_i^j + f_i^\ell + h_i^j + h_i^\ell \right) P_{2,i} & \star & \star \\ \Phi_{21,j\ell} & \Phi_{22,j\ell} & \star \\ \tilde{B}_{w2,j}^T + \tilde{B}_{w2,\ell}^T & 0 & 2I \end{bmatrix} = \Phi_{j\ell} > 0 \qquad (2.63)$$

with

$$\Phi_{21,j\ell} = G_{2,j}^T \tilde{A}_\ell^T + G_{2,\ell}^T \tilde{A}_j^T + Z_j^T \tilde{B}_{u,\ell}^T + Z_\ell^T \tilde{B}_{u,j}^T$$

$$\Phi_{22,j\ell} = G_{2,j} + G_{2,j}^T + G_{2,\ell} + G_{2,\ell}^T - \sum_{i=1}^{N} \left(f_i^j + f_i^\ell \right) P_{2,i}$$

for $j = 1, \ldots, M-1$ and $\ell = j+1, \ldots, M$, and

$$\begin{bmatrix} \sum_{i=1}^{N} f_i^j W_i - \tilde{D}_{2w,j} \tilde{D}_{2w,j}^T & \star \\ G_{2,j}^T \tilde{C}_{2z,j}^T + Z_j^T \tilde{D}_{2u,j}^T & G_{2,j} + G_{2,j}^T - \sum_{i=1}^{N} f_i^j P_{2,i} \end{bmatrix} = \Psi_j > 0 \quad (2.64)$$

for $j = 1, \ldots, M$, and

$$\begin{bmatrix} \sum_{i=1}^{N} \left(f_i^j + f_i^\ell \right) W_i - \tilde{D}_{2w,j} \tilde{D}_{2w,\ell}^T + \tilde{D}_{2w,\ell} \tilde{D}_{2w,j}^T & \star \\ G_{2,j}^T \tilde{C}_{2z,\ell}^T + G_{2,\ell}^T \tilde{C}_{2z,j}^T + Z_j^T \tilde{D}_{2u,\ell}^T + Z_\ell^T \tilde{D}_{2u,j}^T & \Psi_{22,j\ell} \end{bmatrix} = \Psi_{j\ell} > 0 \quad (2.65)$$

with

$$\Psi_{22,j\ell} = G_{2,j} + G_{2,j}^T + G_{2,\ell} + G_{2,\ell}^T - \sum_{i=1}^{N} \left(f_i^j + f_i^\ell \right) P_{2,i}$$

for $j = 1, \ldots, M-1$ and $\ell = j+1, \ldots, M$ where

$$G_{\infty j} = \begin{bmatrix} \sum_{i=1}^{N} f_i^j G_{i,1} & 0 \\ G_{\infty j,2} & G_{\infty j,3} \end{bmatrix}, \quad G_{2j} = \begin{bmatrix} \sum_{i=1}^{N} f_i^j G_{i,1} & 0 \\ G_{2j,2} & G_{2j,3} \end{bmatrix}, \quad and$$

$$Z_j = \begin{bmatrix} \sum_{i=1}^{N} f_i^j Z_{i,1} & 0 \end{bmatrix}, \qquad (2.66)$$

then the parameter-dependent static output feedback gain

$$K(\lambda_k) = \hat{Z}(\lambda_k) \hat{G}(\lambda_k)^{-1} \qquad (2.67)$$

with

$$\hat{Z}(\lambda_k) = \sum_{i=1}^{N} \lambda_i(k) Z_{i,1} \quad and \quad \hat{G}(\lambda_k) = \sum_{i=1}^{N} \lambda_i(k) G_{i,1} \qquad (2.68)$$

stabilizes the system H^∞ with a guaranteed $\mathcal{H}_\infty$ performance bounded by η and the system H^2 with a guaranteed $\mathcal{H}_2$ performance bounded by v given by

$$v^2 = \min_{P_{\infty,i},P_{2,i},G_{i,1},G_{\infty j,2},G_{2j,2},G_{\infty j,3},G_{2j,3},Z_{i,1},W_i} \max_i \text{trace}\{W_i\}. \tag{2.69}$$

The proof for Lemma 2.9 is provided by [13].

Note that, as with any multi-objective controller synthesis, the mixed $\mathcal{H}_2/\mathcal{H}_\infty$ controller synthesis LMIs in Lemma 2.9 can be solved in a few different ways depending on the needs of the control designer. For instance, a controller with the best possible $\mathcal{H}_2$ performance is found with respect to a fixed, predetermined $\mathcal{H}_\infty$ performance η by solving the LMIs to minimize

$$\sum_{i=1}^{N} \text{trace}\{W_i\},$$

while ensuring that $\|H^\infty\|_\infty < \eta$. Likewise, a controller with the best possible $\mathcal{H}_\infty$ performance is found with respect to a fixed, predetermined $\mathcal{H}_2$ performance by first adding the following LMIs to the controller synthesis:

$$\overline{W} - W_i > 0, \quad i = 1, \ldots, N, \tag{2.70}$$

where $\overline{W}$ is selected to provide the desired $\mathcal{H}_2$ performance, and then minimizing η in the $\mathcal{H}_\infty$ LMIs.

Suppose that a control design problem or application had certain system outputs that were required to maintain hard constraints instead of just minimizing a weighted $\mathcal{H}_2$ or $\mathcal{H}_\infty$ performance. This would require that the closed-loop system have a guaranteed ℓ_2 to ℓ_∞ gain, which will be covered in the next chapter.

Chapter 3
Guaranteed $\ell_2 - \ell_\infty$ Gain Control for LPV Systems

This chapter considers the optimal control of polytopic, discrete-time LPV systems with a guaranteed ℓ_2 to ℓ_∞ gain. Additionally, to guarantee robust stability of the closed-loop system under parameter variations, $\mathscr{H}_\infty$ performance criterion is also considered as well. Controllers with a guaranteed ℓ_2 to ℓ_∞ gain and a guaranteed $\mathscr{H}_\infty$ performance (ℓ_2 to ℓ_2 gain) are mixed $\mathscr{H}_2/\mathscr{H}_\infty$ controllers. Normally, $\mathscr{H}_2$ controllers are obtained by considering a quadratic cost function that balances the output performance with the control input needed to achieve that performance. However, to obtain a controller with a guaranteed ℓ_2 to ℓ_∞ gain (closely related to the physical performance constraint), the cost function used in the $\mathscr{H}_2$ control synthesis minimizes the control input subject to maximal singular-value performance constraints on the output. This problem can be efficiently solved by a convex optimization with LMI constraints. The contribution of this chapter is the characterization of the control synthesis LMIs used to obtain an LPV controller with a guaranteed ℓ_2 to ℓ_∞ gain and $\mathscr{H}_\infty$ performance while the control ℓ_2 to ℓ_∞ gain is minimized. A numerical example is presented to demonstrate the effectiveness of the convex optimization.

3.1 Introduction

The design of multi-objective, mixed $\mathscr{H}_2/\mathscr{H}_\infty$ controllers has been a topic of interest for sometime [10, 13, 16, 18, 32, 33, 36, 51, 52]. The goal of using both $\mathscr{H}_2$ and $\mathscr{H}_\infty$ performance criteria is to design a controller which can meet multiple performance objectives. In [10, 51] mixed $\mathscr{H}_2/\mathscr{H}_\infty$ control was introduced by minimizing the $\mathscr{H}_2$ norm of a closed-loop transfer function subject to an $\mathscr{H}_\infty$ norm constraint of another closed-loop transfer function. In [33], mixed $\mathscr{H}_2/\mathscr{H}_\infty$ state-feedback and output-feedback controllers were designed for continuous-time systems by using a convex optimization approach to solve the coupled nonlinear matrix Riccati equations and in [32] a similar approach is used for discrete-time systems. The state-feedback $\mathscr{H}_2/\mathscr{H}_\infty$ design with regional pole placement was addressed by [16] using the LMI

A. P. White et al., *Linear Parameter-Varying Control for Engineering Applications*,
SpringerBriefs in Control, Automation and Robotics,
DOI: 10.1007/978-1-4471-5040-4_3, © The Author(s) 2013

approach. In [36, 52], the LMI approach for multi-objective control synthesis for output-feedback controllers is presented. In [19], an extra instrumental variable was added to the LMI stability conditions to build a parameter dependent Lyapunov function capable of proving the stability of uncertain LTI systems. The new extended LMI conditions in [19] were used in [18] to develop $\mathcal{H}_2$ and $\mathcal{H}_\infty$ LMI conditions for linear state-feedback and output-feedback controllers for uncertain LTI systems. The extended LMI conditions provided by [18] were utilized in [12] and [11] to develop LPV static output feedback controllers that meet $\mathcal{H}_2$ [12] and $\mathcal{H}_\infty$ [11] performance bounds for LTV systems with polytopic uncertainty. The results presented in [12] and [11] were extended in [13] to cover the synthesis of multi-objective $\mathcal{H}_2/\mathcal{H}_\infty$ gain-scheduled output feedback controllers.

Gain scheduling controllers designed using the LPV method have traditionally included $\mathcal{H}_\infty$ performance constraints. This is largely due to the fact that $\mathcal{H}_\infty$ controllers can provide robust stability margins that $\mathcal{H}_2$ controllers cannot provide [72]. However, since the $\mathcal{H}_\infty$ norm is defined as the root-mean-square gain, or ℓ_2 to ℓ_2 gain, from the exogenous input to the regulated output, controllers designed with only $\mathcal{H}_\infty$ performance constraints are not suitable for use when hard constraints on responses or actuator signals must be met.

When hard constraints on responses or actuator signals must be met, a controller with a guaranteed ℓ_2 to ℓ_∞ gain is required, which is a special type of $\mathcal{H}_2$ controller [76]. Recall that for the deterministic interpretation, as discussed in Sect. 2.2, the $\mathcal{H}_2$ norm is given by the ℓ_∞ norm (peak magnitude) of the performance output divided by the ℓ_2 norm (energy) of the bounded ℓ_2 input. A controller with a guaranteed ℓ_2 to ℓ_∞ gain provides strict bounds on the regulated output while minimizing the control input as much as possible. This problem was solved for LTI systems in [76], where it is referred to as the output covariance constraint (OCC) problem. The OCC problem defined in [76] is to find a controller for a given system to minimize the weighted control input cost subject to a set of output constraints. The OCC problem has two interesting interpretations: stochastic and deterministic. The stochastic interpretation is obtained by first assuming that the $\mathcal{H}_2$ exogenous inputs are uncorrelated zero-mean white noises with a given intensity. Then the OCC problem minimizes the weighted control input covariance subject to the output covariance constraints, such that the constraints are interpreted as constraints on the variance of the performance variables. The deterministic interpretation is obtained by assuming that the $\mathcal{H}_2$ exogenous inputs are unknown but belong to a bounded ℓ_2 energy set. Then the OCC problem minimizes the weighted control input while ensuring that the maximum singular values, or ℓ_∞ response, of the regulated outputs are less than the corresponding output constraints. In other words, the OCC problem is the problem of minimizing the weighted sum of worst-case peak values on the control signals subject to the constraints on the worst-case peak values of the performance variables. This interpretation is important in applications where hard constraints on responses or actuator signals cannot be ignored, such as space telescope pointing [75] and machine tool control. For both interpretations, a solution to the OCC control problem results in a controller with a guaranteed ℓ_2 to ℓ_∞ gain.

The main contribution of this chapter is the state-feedback control synthesis LMIs for discrete-time polytopic LPV systems in Sect. 3.3. When these LMIs are satisfied, the optimal state-feedback LPV controller obtained guarantees that for a finite disturbance energy, hard constraints on the regulated output are met. The guaranteed ℓ_2 to ℓ_∞ gain is achieved by modifying the $\mathcal{H}_2$ control synthesis LMIs provided by [13] to minimize the weighted control input cost while ensuring the output covariances meet the performance constraints. Additionally, the $\mathcal{H}_\infty$ LMIs provided by [13] are modified for the state-feedback LPV control synthesis problem to guarantee the robust stability of the closed-loop system under parameter variations.

This chapter is organized as follows. Section 3.2 formulates the mixed $\ell_2 - \ell_\infty / \mathcal{H}_\infty$ control problem to obtain a controller that has a guaranteed ℓ_2 to ℓ_∞ gain. A set of LMIs is presented in Sect. 3.3 which can be used to perform a convex optimization. In Sect. 3.4, a numerical example is presented to illustrate the performance of the algorithm. Conclusions of this work are given in Sect. 3.5.

3.2 Problem Formulation for Mixed $\ell_2 - \ell_\infty / \mathcal{H}_\infty$ Control

Consider the following unweighted, open-loop discrete-time polytopic LPV system

$$
\begin{aligned}
x_p(k+1) &= A_p(\lambda_k)x_p(k) + B_p(\lambda_k)u(k) + D_p(\lambda_k)w_p(k) \\
y_p(k) &= C_p(\lambda_k)x_p(k)
\end{aligned}
\tag{3.1}
$$

where $x_p(k)$ is the state, $u(k)$ is the control input, $w_p(k)$ is an exogenous ℓ_2 disturbance, and $y_p(k)$ is the vector of all dynamic variables of interest. The system matrices $A_p(\lambda_k)$, $B_p(\lambda_k)$, $D_p(\lambda_k)$, and $C_p(\lambda_k)$ belong to a polytope similar to $\mathfrak{D}$ in (2.11), with $A_{p,i}$, $B_{p,i}$, $D_{p,i}$, and $C_{p,i}$ the vertices of the polytope and $\lambda(k) \in \mathbb{R}^N$ the vector of time-varying barycentric coordinates lying in the unit simplex (2.12). For all $k \in \mathbb{Z}_{\geq 0}$, the rate of variation of the parameters (2.26), reproduced here

$$
\Delta\lambda_i(k) = \lambda_i(k+1) - \lambda_i(k), \quad i = 1, \ldots, N,
$$

is assumed to be limited by an a priori bound $b \in [0, 1]$ as in (2.35). The uncertainty domain, where the vector $(\lambda(k), \Delta\lambda(k))^T \in \mathbb{R}^{2N}$ takes values, can be modeled by the compact set Γ_b given in (2.36).

Suppose that we apply to the plant (3.1) a full state feedback stabilizing control law of the form

$$
u(k) = K(\lambda_k)x_p(k).
\tag{3.2}
$$

Then the resulting closed-loop system is

$$
\begin{aligned}
x(k+1) &= \mathscr{A}(\lambda_k)x(k) + \mathscr{B}(\lambda_k)w_2(k) \\
y_p &= C_p(\lambda_k)x(k)
\end{aligned}
\tag{3.3}
$$

where $x = x_p$, $w_2 = w_p$, and

$$\mathscr{A}(\lambda_k) = A_p(\lambda_k) + B_p(\lambda_k)K(\lambda_k),$$
$$\mathscr{B}(\lambda_k) = D_p(\lambda_k).$$

Theorem 3.1. *Consider the asymptotically stable system (3.3). Define the ℓ_2 and ℓ_∞ norms as*

$$\|y_p\|_\infty^2 = \sup_{k\geq 0} y_p^T(k)y_p(k), \quad \|w_2\|_2^2 = \sum_{\ell=0}^\infty w_2^T(\ell)w_2(\ell). \tag{3.4}$$

Then the ℓ_2 to ℓ_∞ gain of (3.3) is

$$\frac{\|y_p\|_\infty^2}{\|w_2\|_2^2} \leq \bar{\sigma}(Y), \tag{3.5}$$

where $Y = C_p(\lambda_k)\bar{P}^\infty C_p(\lambda_k)^T$, $\bar{P}^\infty = \lim_{k\to\infty}\bar{P}(k)$, and $\bar{P}(k)$ is the solution of the time-varying Lyapunov equation

$$\bar{P}(\lambda_{k+1}) = \mathscr{A}(\lambda_k)\bar{P}(\lambda_k)\mathscr{A}(\lambda_k)^T + \mathscr{B}(\lambda_k)\mathscr{B}(\lambda_k)^T. \tag{3.6}$$

A proof for Theorem 3.1. can be obtained using operator theory by following the steps provided by [28].

Suppose that some a priori information about the constraints on the performance of y_p are known such that an output covariance bound $\bar{Y}$ can be constructed. It is the purpose of this chapter to design an LPV state-feedback controller with

$$\|y_p\|_\infty^2 \leq \bar{\sigma}(\bar{Y})\|w_2\|_2^2, \tag{3.7}$$

such that the guaranteed ℓ_2 to ℓ_∞ gain is

$$\sup_{w_2\in\ell_2,w_2\neq 0} \frac{\|y_p\|_\infty^2}{\|w_2\|_2^2} \leq \bar{\sigma}(\bar{Y}). \tag{3.8}$$

Thus we are interested in finding a controller of the form (3.2) that minimizes the (weighted) control energy

$$J_{OCC} = \text{trace}\left\{RK(\lambda_k)\bar{P}(\lambda_k)K^T(\lambda_k)\right\}, \quad R > 0 \tag{3.9}$$

and satisfies the constraint

$$Y(\lambda_k) = C_p(\lambda_k)\bar{P}(\lambda_k)C_p^T(\lambda_k) \leq \bar{Y}. \tag{3.10}$$

Suppose that in addition to meeting the constraint (3.10) while minimizing the control energy (3.9), we also desire the closed-loop system (3.3) to be robust to low-frequency parameter errors, additive plant errors, and parameter variations. Then we must also consider some $\mathscr{H}_\infty$ performance criteria. There exist many different $\mathscr{H}_\infty$ weighting schemes (see [54, 72]) that can be used depending on the desired robustness properties of the closed-loop system.

Once an appropriate $\mathscr{H}_\infty$ weighting scheme and ℓ_2 to ℓ_∞ gain control input weight R in (3.9) are selected, then the following discrete-time polytopic time-varying systems H^∞ (2.49) and H^σ can be constructed:

$$H^\infty := \begin{cases} x(k+1) = A(\lambda_k)x(k) + B_\infty(\lambda_k)w_\infty(k) + B_u(\lambda_k)u(k) \\ z_\infty(k) = C_z(\lambda_k)x(k) + D_w(\lambda_k)w_\infty(k) + D_u(\lambda_k)u(k) \end{cases} \quad (3.11)$$

$$H^\sigma := \begin{cases} x(k+1) = A(\lambda_k)x(k) + B_\sigma(\lambda_k)w_2(k) + B_u(\lambda_k)u(k) \\ y_p(k) = C_p(\lambda_k)x(k) \\ z_\sigma(k) = D_{\sigma u}(\lambda_k)u(k) \end{cases} \quad (3.12)$$

where $x(k) \in \mathbb{R}^n$ is the state, $w_\infty \in \mathbb{R}^{r_\infty}$ and $w_2(k) \in \mathbb{R}^{r_\sigma}$ are the $\mathscr{H}_\infty$ and ℓ_2 to ℓ_∞ gain exogenous inputs, and $u(k) \in \mathbb{R}^m$ is the control input. The outputs $z_\infty(k) \in \mathbb{R}^{p_\infty}$ and $z_\sigma(k) \in \mathbb{R}^{p_\sigma}$ are the weighted system performance outputs for the mixed ℓ_2 - $\ell_\infty/\mathscr{H}_\infty$ control synthesis, while the output $y_p(k) \in \mathbb{R}^c$ contains all variables whose dynamic responses have hard constraints that must be met.

We note at this juncture that while we have set up the mixed ℓ_2 - $\ell_\infty/\mathscr{H}_\infty$ control problem for one ℓ_2 to ℓ_∞ gain constraint and one $\mathscr{H}_\infty$ performance constraint, the framework for the control synthesis problem can be easily extended to include multiple ℓ_2 to ℓ_∞ gains and $\mathscr{H}_\infty$ performance constraints.

3.3 An Algorithm

The problem posed in the previous section is solved by performing a convex optimization over a set of LMIs. The LMIs in this section are an extension of the work presented in [13]. Using the parameterizations (2.25), (2.39), (2.40), and (2.41), the following finite-dimensional LMIs can be solved to obtain a full-state feedback controller (3.2) such that the closed-loop systems for H^σ and H^∞ have guaranteed ℓ_2 to ℓ_∞ gain and $\mathscr{H}_\infty$ norm, respectively.

Theorem 3.2. *Consider the system H^σ, given by (3.12). Assume that the vectors f^j and h^j of Γ_b are given. Given $\overline{Y}$, if there exists, for $i = 1, 2, \ldots, N$, matrices, $G_i \in \mathbb{R}^{n \times n}$ and $Z_i \in \mathbb{R}^{m \times n}$, and symmetric positive-definite matrices $P_{\sigma,i} \in \mathbb{R}^{n \times n}$ and $W_i \in \mathbb{R}^{p_\sigma \times p_\sigma}$ such that*

$$\begin{bmatrix} \hat{P}_{\sigma,j} & \star & \star \\ \tilde{G}_j^T \tilde{A}_j^T + \tilde{Z}_j^T \tilde{B}_{u,j}^T \tilde{G}_j + \tilde{G}_j^T - \tilde{P}_{\sigma,j} & \star \\ \tilde{B}_{\sigma,j}^T & 0 & I \end{bmatrix} = \Phi_j > 0 \qquad (3.13)$$

for $j = 1, 2, \ldots, M$, and

$$\begin{bmatrix} \hat{P}_{\sigma,j} + \hat{P}_{\sigma,\ell} & \star & \star \\ \tilde{G}_j^T \tilde{A}_\ell^T + \tilde{G}_\ell^T \tilde{A}_j^T + \tilde{Z}_j^T \tilde{B}_{u,\ell}^T + \tilde{Z}_\ell^T \tilde{B}_{u,j}^T \tilde{G}_j + \tilde{G}_j^T + \tilde{G}_\ell + \tilde{G}_\ell^T - \tilde{P}_{\sigma,j} - \tilde{P}_{\sigma,\ell} & \star \\ \tilde{B}_{\sigma,j}^T + \tilde{B}_{\sigma,\ell}^T & 0 & 2I \end{bmatrix} = \Phi_{j\ell} > 0$$

$$(3.14)$$

for $j = 1, \ldots, M - 1$ and $\ell = j + 1, \ldots, M$, and

$$\begin{bmatrix} \tilde{W}_j & \star \\ \tilde{Z}_j^T \tilde{D}_{2u,j}^T \tilde{G}_j + \tilde{G}_j^T - \tilde{P}_{\sigma,j} \end{bmatrix} = \Psi_j > 0 \qquad (3.15)$$

for $j = 1, 2, \ldots, M$ and

$$\begin{bmatrix} \tilde{W}_j + \tilde{W}_\ell & \star \\ \tilde{Z}_j^T \tilde{D}_{2u,\ell}^T + \tilde{Z}_\ell^T \tilde{D}_{2u,j}^T \tilde{G}_j + \tilde{G}_j^T + \tilde{G}_\ell + \tilde{G}_\ell^T - \tilde{P}_{\sigma,j} - \tilde{P}_{\sigma,\ell} \end{bmatrix} = \Psi_{j\ell} > 0$$

$$(3.16)$$

for $j = 1, \ldots, M - 1$ and $\ell = j + 1, \ldots, M$ and

$$\overline{Y} - C_{p,i} P_{\sigma,i} C_{p,i}^T \geq 0, \quad i = 1, 2, \ldots, N, \qquad (3.17)$$

with

$$\hat{P}_{\sigma,j} = \sum_{i=1}^{N} \left(f_i^j + h_i^j \right) P_{\sigma,i}, \quad \tilde{P}_{\sigma,j} = \sum_{i=1}^{N} f_i^j P_{\sigma,i},$$

$$\tilde{G}_j = \sum_{i=1}^{N} f_i^j G_i, \quad \tilde{Z}_j = \sum_{i=1}^{N} f_i^j Z_i, \quad and \quad \tilde{W}_j = \sum_{i=1}^{N} f_i^j W_i,$$

then the parameter-dependent full state feedback gain

$$K(\lambda_k) = \hat{Z}(\lambda_k) \hat{G}(\lambda_k)^{-1} \qquad (3.18)$$

with

$$\hat{Z}(\lambda_k) = \sum_{i=1}^{N} \lambda_i(k) Z_i \quad and \quad \hat{G}(\lambda_k) = \sum_{i=1}^{N} \lambda_i(k) G_i \qquad (3.19)$$

stabilizes the the system H^σ with a guaranteed (weighted) control energy bounded by $\overline{J}_{\text{OCC}}$ given by

$$\overline{J}_{\text{OCC}} = \min_{P_{\infty,i},P_{\sigma,i},G_i,Z_i,W_i} \max_i \text{trace}\{W_i\}$$

$$\geq \text{trace}\{RK(\lambda)P_\sigma(\lambda)K^T(\lambda)\} = J_{\text{OCC}} \qquad (3.20)$$

while also ensuring that the output constraint (3.10) is satisfied. In addition, consider the system H^∞, given by (2.49). If there exist, for $i = 1, 2, \ldots, N$, symmetric positive-definite matrices $P_{\infty,i} \in \mathbb{R}^{n \times n}$ such that

$$\begin{bmatrix} \hat{P}_{\infty,j} & \star & & \star & \star \\ \tilde{G}_j^T \tilde{A}_j^T + \tilde{Z}_j^T \tilde{B}_{u,j}^T & \tilde{G}_j + \tilde{G}_j^T - \tilde{P}_{\infty,j} & \star & \star \\ \tilde{B}_{\infty,j}^T & 0 & \eta I & \star \\ 0 & \tilde{C}_{z,j}\tilde{G}_j + \tilde{D}_{u,j}\tilde{Z}_j & \tilde{D}_{w,j} & \eta I \end{bmatrix} = \Theta_j > 0 \qquad (3.21)$$

for $j = 1, 2, \ldots, M$ and

$$\begin{bmatrix} \hat{P}_{\infty,j} + \hat{P}_{\infty,\ell} & \star & & \star & \star \\ \Theta_{21,j\ell} & \Theta_{22,j\ell} & \star & \star \\ \tilde{B}_{\infty,j}^T + \tilde{B}_{\infty,\ell}^T & 0 & 2\eta I & \star \\ 0 & \Theta_{42,j\ell} & \tilde{D}_{w,j} + \tilde{D}_{w,\ell} & 2\eta I \end{bmatrix} = \Theta_{j\ell} > 0 \qquad (3.22)$$

with

$$\Theta_{21,j\ell} = \tilde{G}_j^T \tilde{A}_\ell^T + \tilde{G}_\ell^T \tilde{A}_j^T + \tilde{Z}_j^T \tilde{B}_{u,\ell}^T + \tilde{Z}_\ell^T \tilde{B}_{u,j}^T$$

$$\Theta_{22,j\ell} = \tilde{G}_j + \tilde{G}_j^T + \tilde{G}_\ell + \tilde{G}_\ell^T - \tilde{P}_{\infty,j} - \tilde{P}_{\infty,\ell}$$

$$\Theta_{42,j\ell} = \tilde{C}_{z,j}\tilde{G}_\ell + \tilde{C}_{z,\ell}\tilde{G}_j + \tilde{D}_{u,j}\tilde{Z}_\ell + \tilde{D}_{u,\ell}\tilde{Z}_j$$

for $j = 1, 2, \ldots, M - 1$ and $\ell = j + 1, \ldots, M$, where

$$\hat{P}_{\infty,j} = \sum_{i=1}^{N} \left(f_i^j + h_i^j\right) P_{\infty,i}, \quad \tilde{P}_{\infty,j} = \sum_{i=1}^{N} f_i^j P_{\infty,i},$$

then the parameter-dependent full-state feedback gain $K(\lambda_k)$ given by (3.18) also stabilizes the system H^∞ with a guaranteed $\mathcal{H}_\infty$ performance bounded by η.

Proof. The sketch of the proof is provided as follows. The following properties are a consequence of applying Lemma 2.9:

- The system H^∞ is stabilized with a guaranteed $\mathcal{H}_\infty$ performance bounded by η when the LMIs (3.21) and (3.22) are satisfied.

- The system H^σ is stabilized with a guaranteed (weighted) control energy bounded by $\bar{J}_{OCC}$ (3.20) when the LMIs (3.13), (3.14), (3.15), and (3.16) are satisfied.

However, the fact that the output constraint (3.10) is satisfied when the LMI (3.17) is satisfied for $i = 1, 2, \ldots, N$ follows from the LMI constraint

$$\bar{Y} - C_{p,i} P_{\sigma,i} C_{p,i}^T \geq 0, \quad i = 1, 2, \ldots, N.$$

Since the LMIs (3.13), (3.14), (3.15), and (3.16) are all required to be positive-definite, from [13] it can be shown that

$$P_\sigma(\lambda_k) = \sum_{i=1}^N \lambda_i(k) P_{\sigma,i} > \bar{P}(\lambda_k), \quad \forall k \geq 0,$$

where $\bar{P}(\alpha_k)$ is the controllability Gramian satisfying (3.6). Thus, it is also true that

$$\bar{Y} - C_p(\lambda_k) \bar{P}(\lambda_k) C_p^T(\lambda_k) \geq 0$$

such that

$$Y(\lambda_k) = C_p(\lambda_k) \bar{P}(\lambda_k) C_p^T(\lambda_k) \leq \bar{Y}.$$

$\square$

3.4 Numerical Example

Consider the discrete-time LPV system (originally used in [20], and later used in [4, 18]).

$$x_p(k+1) = \underbrace{\begin{bmatrix} 2+\delta_1 & 0 & 1 \\ 1 & 0.5 & 0 \\ 0 & 1 & -0.5 \end{bmatrix}}_{A_p(\delta_1)} x_p(k) + \underbrace{\begin{bmatrix} 1+\delta_2 \\ 0 \\ 0 \end{bmatrix}}_{B_p(\delta_2)} u(k) + \underbrace{\begin{bmatrix} 0 \\ 1 \\ 0 \end{bmatrix}}_{D_p} w_p(k)$$

$$y_p(k) = \underbrace{\begin{bmatrix} 1 & 0 & 0 \\ 0 & 1 & 0 \\ 0 & 0 & 1 \end{bmatrix}}_{C_p} x_p(k) \tag{3.23}$$

where δ_i, $i = 1, 2$ are the time-varying parameters, which are assumed to have the following parameter variation bounds:

$$\delta_1 \in [-1, 1], \quad \text{and} \quad \delta_2 \in [-0.5, 0.5]. \tag{3.24}$$

The discrete-time LPV system (3.23) is converted to the discrete-time polytopic LPV system (3.1) by solving $A_p(\delta_1)$ and $B_p(\delta_2)$ at the vertices of the parameter space polytope of δ_1 and δ_2. The exogenous ℓ_2 disturbance w_p is a scalar and the performance variable y_p has three components. The weighting matrice required in (3.9) is taken to be

$$R = 1.$$

In the following, we consider two different ℓ_2 to ℓ_∞ gain designs. The designs differ in the grouping of the performance variables inside of y_p used to define the constraints (3.10). The constraints for each design are given as follows:

$$\textbf{Design 1:} \quad Y \leq 1.85 \times I_3, \qquad\qquad (3.25)$$
$$\textbf{Design 2:} \quad Y_1 \leq 1.85, \quad Y_2 \leq 1.85 \times I_2, \qquad (3.26)$$

where for design 1, Y denotes the (3×3) output covariance matrix corresponding to the all performance outputs in y_p grouped together. In design 2, Y_1 denotes the (1×1) output variance corresponding to the first performance output of y_p and Y_2 denotes the (2×2) output covariance matrix corresponding to the second and third performance outputs grouped together.

For each design, to enhance the robustness of the closed-loop system using the controller $K(\lambda_k)$ with respect to uncertainty in the measurements of the time-varying parameters δ_1 and δ_2, the closed-loop $\mathscr{H}_\infty$ norms of the transfer functions of some appropriately defined extra inputs and outputs that 'pull out' [18, 20] the uncertain parameters are bounded. Specifically, the following H^∞ system is defined:

$$H^\infty = \begin{cases} x(k+1) = A_p(\delta_1)x(k) + B_p(\delta_2)u(k) + \begin{bmatrix} 1 \\ 0 \\ 0 \end{bmatrix} w_{\infty,1}(k) + \begin{bmatrix} 1 \\ 0 \\ 0 \end{bmatrix} w_{\infty,2}(k) \\[12pt] z_{\infty,1}(k) = \begin{bmatrix} 1 & 0 & 0 \end{bmatrix} x(k) \\ z_{\infty,2}(k) = u(k) \end{cases}$$

$$(3.27)$$

so that the robustness requirement is given by

$$\| H_{z_{\infty,i} w_{\infty,i}}(\alpha) \|_\infty < \eta = 100, \quad i = 1, 2, \qquad (3.28)$$

where η defines the robustness level. Note that the notation used here, specifically $w_{\infty,1}(k)$ and $w_{\infty,2}(k)$ with the same input matrix, was selected to match what is found in the literature [18, 20].

For each of the $\ell_2 - \ell_\infty$ designs (3.25)–(3.26), the LMIs in Sect. 3.3 are programmed into MATLAB using the LMI parser YALMIP [35] and solved with SeDuMi [56] to minimize the cost function (3.20). As shown in Figs. 3.1 and 3.3a, each design is feasible and the achieved covariance bound is tight with the design bound in at least one dimension. The constraint in design 1 ensures that the covariance bound ellipsoid of Y remains inside of the sphere displayed in Fig. 3.1a. Side views

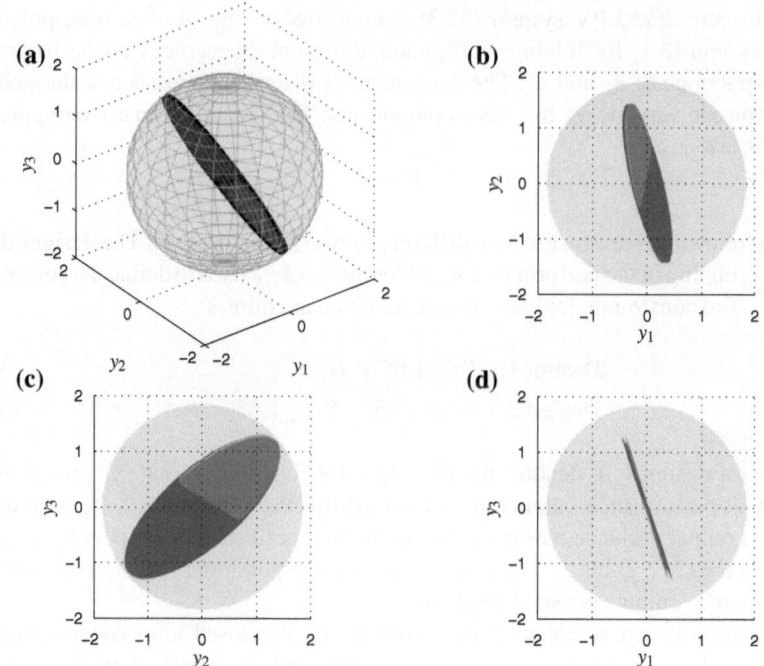

Fig. 3.1 Design 1: The covariance bound Y achieved compared to the constraint (3.25)

of the covariance bound Y are displayed in Fig. 3.1b–d. As displayed in Fig. 3.1c, the output covariance Y is tight with the bound in the y_2-y_3 plane.

For design 2, the constraints ensure that the variance of the first output of y_p will be below 1.85 and the covariance bound of second and third outputs of y_p will remain inside of the circle in Fig. 3.3a. The dashed ellipses in Fig. 3.3a are the obtained output covariances at each of the vertices for $i = 1, \ldots, 4$, and as shown they are tight with the bound.

To test the performance of each design, we simulate each of the controllers with a positive impulse ($\mathfrak{I}_1$) followed by a negative impulse ($\mathfrak{I}_2$) as displayed in Fig. 3.4a. To see the effect of the time-varying parameters, the parameters δ_1 and δ_2 are varied as displayed in Fig. 3.4b. The values used to compute the controller at each time step k are the noisy measurements displayed with a gray dashed line. The response to the ℓ_2 disturbance $w_p(k)$ for design 1 is displayed in Fig. 3.2. The response in Fig. 3.2 is plotted inside of the ℓ_∞ norm constraint (the square root of the covariance bound) sphere and the achieved ℓ_∞ norm bound ellipsoid. In Fig. 3.3b, the response of design 2 is plotted inside of the ℓ_∞ norm constraint circle and the achieved ℓ_∞ norm bound ellipse. The path of the response, with respect to each of the impulses ($\mathfrak{I}_1$) and ($\mathfrak{I}_2$), is also displayed in Fig. 3.3b. As shown in Figs. 3.2 and 3.3b, the response for each design stays inside of the ℓ_∞ bound.

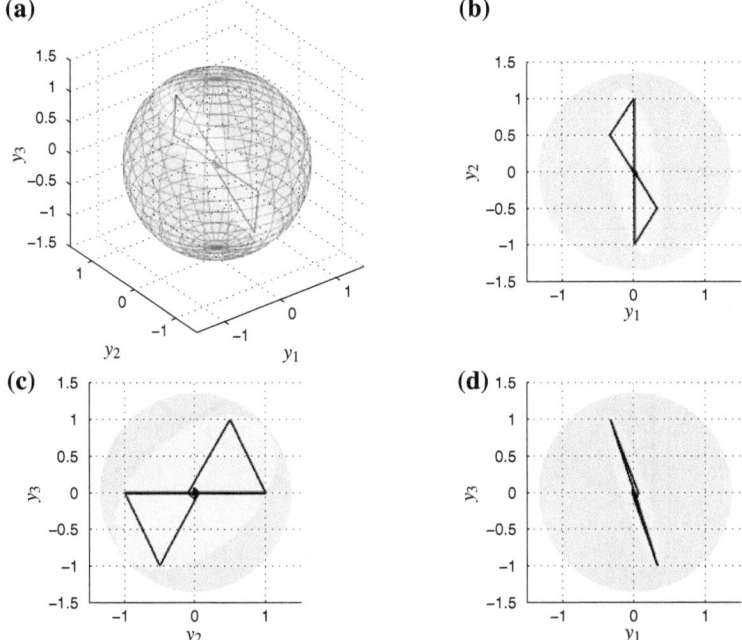

Fig. 3.2 Design 1: The output response of y_1, y_2, and y_3 plotted against each other for design 1 simulated with a positive (I_1) and negative (I_2) impulse function and compared with the ℓ_∞ norm bound

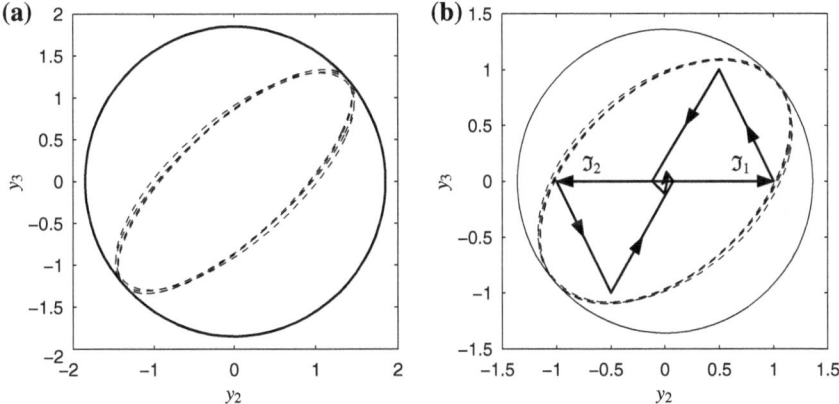

Fig. 3.3 Design 2: a The covariance bound Y_2 achieved compared to the second constraint in (3.26). **b** The output response of y_2 plotted against y_3 for design 2 simulated with a positive and negative impulse function and compared with the ℓ_∞ norm bound

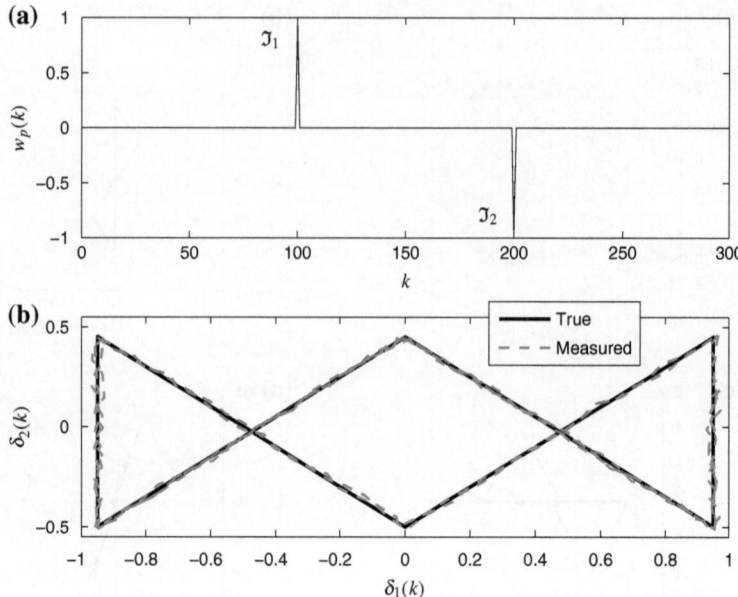

Fig. 3.4 The ℓ_2 disturbance (**a**) and the parameter variation (**b**) used to simulate each controller design

3.5 Conclusion

In this chapter, we have considered the state-feedback control of polytopic, discrete-time LPV systems with a guaranteed ℓ_2 to ℓ_∞ gain. $\mathcal{H}_\infty$ performance criteria is also considered to guarantee the robust stability of the closed-loop system. To obtain the gain-scheduling controller with guaranteed $\ell_2 - \ell_\infty$ and $\ell_2 - \ell_2$ gains, a set of LMIs are presented in Sect. 3.3. Finally, the LMIs are solved to obtain state-feedback gain-scheduling controllers for two different designs for a numerical example.

Chapter 4
Gain-Scheduling Control of Port-Fuel-Injection Processes

In this chapter, an LPV design example [61, 62] that demonstrates how to design gain-scheduling proportional-integral (PI) and proportional-integral-derivative (PID) controllers using the LPV methods from Chap. 2 is presented. First, physics-based modeling is used to create an event-based sampled discrete-time linear system representing a port-fuel-injection process based on wall-wetting dynamics, which is then formulated as an LPV system. Then a control strategy is developed and relevant control structures are appended onto the LPV system to produce the generalized LPV plant. The generalized LPV plant is then used with the $\mathscr{H}_\infty$ LPV controller synthesis presented in Sect. 2.3.1 to synthesize the LPV controller. To validate the LPV controller, first a simple simulation was performed. To further validate the LPV controller, a hardware-in-the-loop (HIL) simulation is performed with a mixed mean-value and crank-based engine model.

4.1 Introduction

Increasing concerns about global climate change and ever-increasing demands on fossil fuel capacity call for reduced emissions and improved fuel economy. Port-fuel-injection (PFI) fuel systems are widely used in vehicles today; however, direct-injection (DI) fuel systems have also been introduced in markets globally. To improve the full load performance of DI engines at high speed, Toyota introduced an engine with a stoichiometric DI system with a DI injector and an intake port injector for each cylinder (see [31]). The use of gasoline PFI and ethanol DI dual-fuel system to substantially increase gasoline engine efficiency is described by [30]. This shows that with the introduction of DI fuel systems for the internal combustion engine, PFI fuel systems will remain part of the engine fuel system for improved engine performance, which is the main motivation for revisiting the air-to-fuel (A/F) ratio control problem for a PFI fuel system.

A. P. White et al., *Linear Parameter-Varying Control for Engineering Applications*,
SpringerBriefs in Control, Automation and Robotics,
DOI: 10.1007/978-1-4471-5040-4_4, © The Author(s) 2013

There have been several fuel control strategies developed for internal combustion engines to improve efficiency and exhaust emissions. A key development in the evolution was the introduction of a closed-loop fuel injection control algorithm [50], followed by the linear quadratic control method [15], and an optimal control and Kalman filtering design [47]. Specific applications of A/F ratio control based on observer measurements in the intake manifold were developed by [9]. Another approach was based on measurements of exhaust gas A/F ratio measured by the oxygen sensor and on the throttle position [43]. Choi et al. [17] also developed a nonlinear sliding mode control of A/F ratio based on the oxygen sensor feedback. Continuing research efforts of A/F ratio control include adaptive approaches [59, 69], observer-based controllers [46], H_∞ controllers [37], model predictive controllers [39], sliding mode controllers [44], and LPV controllers [20, 71, 77]. Conventional A/F ratio control for automobiles uses both closed-loop feedback and feedforward control to have good steady state and fast transient responses.

For a spark-ignited engine equipped with a port-fuel-injection system, the wall-wetting dynamics are commonly used to model the fuel injection process; and the wall-wetting effects are compensated on the basis of simple LTI models that are tuned and calibrated through engine dynamometer and vehicle tests. These models are quite effective for an engine operated at steady state or slow transition conditions, but are difficult to use at fast transient and other special operational conditions, for instance, during engine cold start. One of the approaches to model the wall-wetting dynamics during engine cold start is to describe it using a family of linear models to approximate the system dynamics at a given engine coolant temperature, speed, and load conditions, that is, to translate the fuel system model into an LPV system.

As stated earlier, the use of LPV modeling to control the A/F ratio of a port-fuel-injection system has been reported by [26, 71, 77]. In [77], a continuous-time, LPV model is developed considering only engine speed as a time-varying parameter. Due to the simplicity of the model used, the issue of engine cold start is not addressed. Furthermore, the control synthesis method used in [77] relies on gridding the parameter space at a finite number of grid points. In [71], a large variable time delay is present in the A/F ratio control loop for a lean burn spark ignition engine. LPV control methods are used to compensate for the variable time delay. In [26], a discrete-time, LPV model is developed with manifold absolute pressure, exhaust valve closing, and inlet valve opening as the time-varying parameters. However, only manifold absolute pressure is used as a scheduling parameter in the gain-scheduling control that is synthesized. Also, [26] does not address the issue of engine cold start. Additionally, all LPV control synthesis methods used by [26] are based in continuous time, relying on Tustin's (bilinear) transformation to convert the discrete-time system into a continuous-time system, thus fixing the engine speed and sampling rate of the discrete-time system. In contrast to all these efforts, in this chapter an event-based, gain-scheduling controller for an event-based, discrete-time LPV system with wall-wetting parameters and engine speed as time-varying parameters is designed. To cope with practical situations, the discrete-time LPV control synthesis method in Lemma 2.7 is used to develop the event-based, gain-scheduling controller. An affine LPV model including the feedforward control dynamics is obtained. Gain-Scheduling controllers have been synthesized to guarantee the robust stability and performance of the affine LPV model.

The contribution of this chapter is as follows. First, an event-based, discrete-time LPV model for the wall-wetting and oxygen sensor dynamics with wall-wetting parameters and engine speed as scheduling variables is developed. Then an event-based, gain-scheduling controller for the derived LPV model is designed. To cope with practical situations, the discrete-time LPV control synthesis method given by [11] is used to develop the event-based, gain-scheduling controller.

The control structures used in this study are proportional-integral (PI) and proportional-integral-derivative (PID). PI controllers are widely used in industry since they are well understood by field control engineers. The PI gains are often calibrated in field tests for the best performance as functions of system operational conditions. However, the system stability and performance are not guaranteed for all time-varying parameters. Therefore, LPV techniques proposed in this chapter are applied to design gain-scheduling PI controllers for guaranteed stability and performance for all time-varying parameters. Furthermore, the addition of derivative control to a PI controller adds an extra layer of complexity. The design of a PID controller at a single operating point can be a difficult iterative procedure, which would make calibrating PID gains as functions of system operational conditions very time-consuming. However, designing a gain-scheduling PID controller using LPV techniques provided in this chapter is as simple as adding a derivative channel to the control input. The ability to design either a gain-scheduling PI or PID controller with guaranteed stability and performance in one shot without requiring hours of calibration is expected to be well received by industrial control engineers.

The process of designing an LPV controller for any automotive application is depicted in Fig. 4.1. Due to the complexity of internal combustion engines, designing controllers for specific engine systems using an entire engine model is extremely difficult if even possible. Therefore, to design a controller for a specific engine subsystem, first a physics-based simplified model is developed to represent the engine subsystem. After the varying parameters are identified, the physics-based model can be transformed into an LPV model. LPV controller design can then be carried

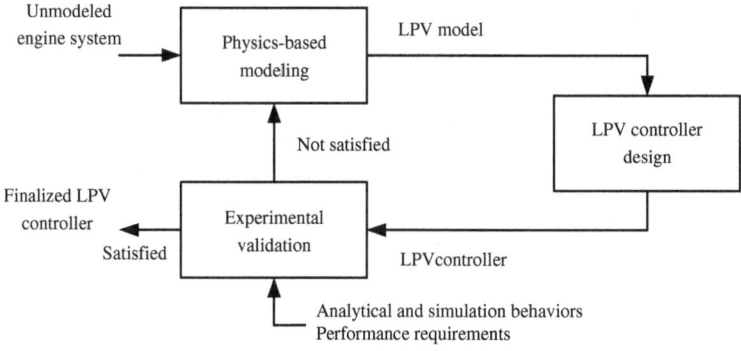

Fig. 4.1 Flowchart of the design and validation process of an LPV controller

out on the LPV model to develop an LPV controller. Once the LPV controller is obtained it must be tested on the original engine to ensure that it meets all stability and performance requirements. A cost-effective way of validating developed LPV controllers is to implement them in a rapid prototyping real-time control systems and validate them through hardware-in-the-loop (HIL) simulations.

In this chapter, we first develop a physics-based model for the port-fuel-injection process based on the wall-wetting dynamics and formulate it as an LPV system. The system parameters used in the engine fuel system model are engine speed, temperature, and load. These system parameters can be obtained in real-time through physical or virtual sensors. A gain-scheduling controller is then obtained for the derived LPV system based on the numerically efficient convex optimization (or LMI) techniques. To validate the gain-scheduling PI and PID controllers, HIL simulations were performed using a mixed mean-value and crank-based engine model [68].

This chapter is organized as follows. The models and the modeling techniques used are given in Sect. 4.2. The design of the gain-scheduling controller in Sect. 4.3 is covered by first introducing the control strategy in Sect. 4.3.1. Then the feed-forward compensated generalized plant is developed in Sect. 4.3.2 and its first-order Taylor series expansion is computed in Sect. 4.3.3. Next, the measurement for control is elaborated in Sect. 4.3.4. The gain-scheduling synthesis problem is stated in Sect. 4.3.5. In Sect. 4.3.6, the augmented LPV plant obtained in Sect. 4.3.4 is converted into a polytopic time-varying system, which is an LPV system with a polytopic dependency on a scheduling parameter that takes values in the unit-simplex, so that the gain-scheduling controller synthesis technique reviewed in Sect. 2.3.1 given by [11] can be performed. For comparison, an LTI feedback $\mathcal{H}_\infty$ controller is designed in Sect. 4.4 using the nominal parameters. Simulation results from three separate engine operating conditions are presented in Sect. 4.5. Next, an HIL simulation setup is introduced in Sect. 4.6 and components of the mixed mean-value and crank-based engine model [68] are reviewed. In Sect. 4.7 the results from the HIL simulations are presented. The concluding remarks are given in the final section.

4.2 Event-Based Discrete-Time System Modeling

In this section, the dynamics of the plant (Fig. 4.2) will be carefully explained and modeled to develop a control-oriented LPV model. The plant given in Fig. 4.2 shows the port-fuel-injection process for a single cylinder engine. However, the methods used in this chapter can be extended to a multiple cylinder engine by using the individual cylinder fuel–gas ratio estimation method developed by [57].

4.2.1 Sampling Period of the Event-Based Discrete-Time System

The discrete-time linear system is obtained by event-based sampling of the port-fuel-injection process; hence the sampling time of this discrete-time system is the period of an engine cycle,

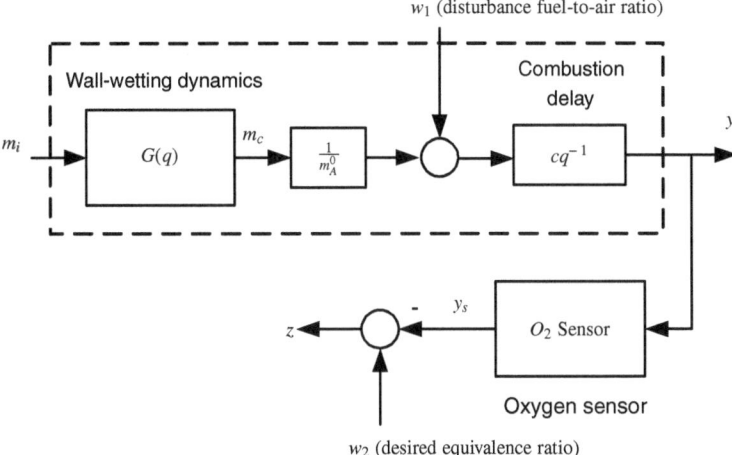

Fig. 4.2 The block diagram of the port-fuel-injection process and sensor dynamics

$$t_s = \frac{1 \text{ min.}}{N_e \text{ rev.}} \left(\frac{60 \text{ sec.}}{1 \text{ min.}} \right) \left(\frac{2 \text{ rev.}}{1 \text{ cycle}} \right) = \frac{120 \text{ sec.}}{N_e \text{ cycle}}, \tag{4.1}$$

where N_e represents the engine speed in revolutions per minute (rpm) (see general engine modeling techniques in [7]).

4.2.2 Dynamics of the Port-Fuel-Injection Process

The wall-wetting dynamics can be described as follows:

$$
\begin{aligned}
m_w(k) &= (1 - \alpha_k)m_w(k - 1) + (1 - \beta_k)m_i(k), \\
m_c(k) &= \alpha_k m_w(k - 1) + \beta_k m_i(k),
\end{aligned}
\tag{4.2}
$$

where $k \in \mathbb{Z}_{\geq 0}$, and m_w, m_c, and m_i denote the amount of fuel, on the wall, in the cylinder, and injected, respectively. The coefficients $\alpha \in [0, 1]$, and $\beta \in [0, 1]$, are the ratios of the fuel delivered from the wall to the cylinder, and of the fuel entering the cylinder from injection, respectively. For notational simplicity, α_k and β_k will be used to denote the wall-wetting parameters at time k, such that $\alpha_k = \alpha(k)$ and $\beta_k = \beta(k)$. These values can be estimated online through an available set of engine sensors, which allows application of gain-scheduling control to the plant. Using the discrete-time dynamics in (4.2), the transfer function $G(q)$ from m_i to m_c is

$$G(q) := \frac{m_c(k)}{m_i(k)} = \frac{\beta_k + (\alpha_k - \beta_k)q^{-1}}{1 - (1 - \alpha_k)q^{-1}}, \tag{4.3}$$

where q is the *forward shift operator* that satisfies $qu(k) = u(k+1)$. The dotted box in the block diagram in Fig. 4.2 illustrates the fuel-injection process. The output of $G(q)$ is the input to the gain block of $\frac{1}{m_A^0}$, which is the nominal value of the inverse of the mass of air trapped in the cylinder m_A. The signal w_1 represents the deviation $\left(\frac{m_c}{m_A} - \frac{m_c^0}{m_A^0} \right)$, which will be treated as a disturbance. Another constant gain factor $c = 14.6$ in Fig. 4.2 is the value for the air-to-fuel-ratio at stoichiometric. After the combustion delay block the equivalence ratio y is generated. The diagram of the transfer function from the amount of fuel injected m_i and the disturbance w_1 to the equivalence ratio y (inverse of normalized air-to-fuel ratio) is shown in the dotted box in Fig. 4.2.

4.2.3 Dynamics of the Oxygen Sensor

To measure y, a UEGO (universal exhaust gas oxygen) sensor is placed in the exhaust manifold at some distance downstream from the exhaust valve. Notice that the continuous-time dynamics and delays will change in the event-based, discrete-time system according to the speed of the engine (or the sampling time). Therefore, the objective of this section is to obtain oxygen sensor dynamics in the form of the finite dimensional, event-based, discrete-time LPV system. Finite dimensionality is required for the applicability of most LPV controller design techniques and the controller design method which will be presented in Sect. 4.3. To this end, in general, one can approximate the continuous-time system with a delay by a finite dimensional event-based, discrete-time LPV system in any standard method. To illustrate this procedure, we demonstrate how we approximate the oxygen sensor dynamics by Taylor series approximation in which the approximation error can be minimized by increasing the order of the Taylor series approximation.

The dynamics of the oxygen sensor are modeled as a first-order sensor time delay coupled with the transport delay of the exhaust gas mixture. The transport delay, $T_D = \frac{d}{N_e}$, of the exhaust gas mixture is both a function of the oxygen sensor placement, which determines the constant d, and the engine speed, N_e. The combined transfer function in the continuous time domain is

$$y_s(s) = \frac{\exp(-T_D s)}{T_{O_2} s + 1} y(s), \tag{4.4}$$

where y_s is the equivalence ratio measured by the sensor and T_{O_2} is the time constant of the oxygen sensor. Since the delay $T_D \in [\frac{d}{N_e}, \frac{d}{N_e}]$ is small, (4.4) can be approximated by the second-order system

$$y_s(s) = \frac{1}{T_D s + 1} \frac{1}{T_{O_2} s + 1} y(s),$$

which has the state–space representation

$$\dot{x}_{O_2} = \underbrace{\begin{bmatrix} -\frac{1}{T_D} & \frac{1}{T_D} \\ 0 & -\frac{1}{T_{O_2}} \end{bmatrix}}_{=:A_{O_2}} x_{O_2} + \underbrace{\begin{bmatrix} 0 \\ \frac{1}{T_{O_2}} \end{bmatrix}}_{=:B_{O_2}} y,$$

$$y_s = \underbrace{\begin{bmatrix} 1 & 0 \end{bmatrix}}_{=:C_{O_2}} x_{O_2}. \tag{4.5}$$

Using t_s as the sampling rate, the corresponding discrete system of (4.5) is

$$x_{O_2}(k+1) = A_{O_2d}x_{O_2}(k) + B_{O_2d}y(k),$$
$$y_s(k) = C_{O_2d}x_{O_2}(k), \tag{4.6}$$

where, due to the invertibility of the matrix A_{O_2} in (4.5),

$$A_{O_2d} = \exp(A_{O_2}t_s),$$

$$B_{O_2d} = \left(\int_0^{t_s} \exp(A_{O_2}\tau)d\tau\right) B_{O_2} = A_{O_2}^{-1}(A_{O_2d} - I)B_{O_2},$$

$$C_{O_2d} = C_{O_2}.$$

Since both T_D and t_s are functions of engine speed, N_e, naturally A_{O_2d} and B_{O_2d} are as well. To capture this parameter variation, the matrices A_{O_2d} and B_{O_2d} are now computed for a given transport delay of $T_D = \frac{80}{N_e}$. To solve for A_{O_2d}, first A_{O_2} is multiplied by t_s

$$A_{O_2}t_s = \begin{bmatrix} -\frac{N_e}{80} & \frac{N_e}{80} \\ 0 & -\frac{1}{T_{O_2}} \end{bmatrix} \frac{120}{N_e}$$

$$= \begin{bmatrix} -\frac{3}{2} & \frac{3}{2} \\ 0 & -\frac{120}{T_{O_2}N_e} \end{bmatrix}. \tag{4.7}$$

Next, the matrix exponent of $A_{O_2}t_s$ is computed, which gives

$$A_{O_2d} = \begin{bmatrix} \exp\left(-\frac{3}{2}\right) & p_1(N_e) \\ 0 & p_2(N_e) \end{bmatrix}, \tag{4.8}$$

where

$$p_1(N_e) = \frac{-\frac{3}{2}\left(\exp\left(-\frac{3}{2}\right) - \exp\left(-\frac{120}{T_{O_2}N_e}\right)\right)}{-\frac{120}{T_{O_2}N_e} + \frac{3}{2}}, \tag{4.9a}$$

$$p_2(N_e) = \exp\left(-\frac{120}{T_{O_2}N_e}\right). \tag{4.9b}$$

To represent the parameter variation in A_{O_2d}, a fourth-order Taylor series approximation of $p_1(N_e)$ and $p_2(N_e)$ is used. To ensure that the coefficients of the Taylor series approximations of $p_1(N_e)$ and $p_2(N_e)$ are numerically stable with respect to the condition number [58], $\frac{1}{N_e} \in [\frac{1}{\overline{N_e}}, \frac{1}{\underline{N_e}}]$ is normalized to γ in the following way:

$$\gamma = \frac{\frac{1}{N_e} - \frac{1}{N_{e,0}}}{\frac{1}{N_e} + \frac{1}{N_{e,0}}} \quad \text{where} \quad \frac{1}{N_{e,0}} = \frac{\frac{1}{\overline{N_e}} + \frac{1}{\underline{N_e}}}{2}. \tag{4.10}$$

Solving (4.10) for $\frac{1}{N_e}$, and substituting into (4.9a) and (4.9b), $p_1(\gamma)$ and $p_2(\gamma)$ are found to be

$$p_1(\gamma) = \frac{-\frac{3}{2}\left(\exp\left(-\frac{3}{2}\right) - \exp\left(-\frac{120}{T_{O_2}N_{e,0}}\left(\frac{1+\gamma}{1-\gamma}\right)\right)\right)}{-\frac{120}{T_{O_2}N_{e,0}}\left(\frac{1+\gamma}{1-\gamma}\right) + \frac{3}{2}}, \tag{4.11a}$$

$$p_2(\gamma) = \exp\left(-\frac{120}{T_{O_2}N_{e,0}}\left(\frac{1+\gamma}{1-\gamma}\right)\right). \tag{4.11b}$$

Finally, the forth-order Taylor series approximation of A_{O_2d} is represented with the following lower LFT:

$$A_{O_2d} = \mathscr{F}_\ell(M_{A_{O_2d}}, \gamma I_4) \tag{4.12}$$

where

$$M_{A_{O_2d}} = \left[\begin{array}{c:cccccc}
\exp\left(-\frac{3}{2}\right) & a_0 & a_1 & a_2 & a_3 & a_4 \\
0 & b_0 & b_1 & b_2 & b_3 & b_4 \\ \hdashline
0 & 1 & 0 & 0 & 0 & 0 \\
0 & 0 & 1 & 0 & 0 & 0 \\
0 & 0 & 0 & 1 & 0 & 0 \\
0 & 0 & 0 & 0 & 1 & 0
\end{array}\right] \tag{4.13}$$

and

$$a_1 = \frac{1}{1!}\frac{d^1 p_1(0)}{d\gamma^1}, \quad a_2 = \frac{1}{2!}\frac{d^2 p_1(0)}{d\gamma^2}, \quad a_3 = \frac{1}{3!}\frac{d^3 p_1(0)}{d\gamma^3}, \quad a_4 = \frac{1}{4!}\frac{d^4 p_1(0)}{d\gamma^4},$$

$$b_1 = \frac{1}{1!}\frac{d^1 p_2(0)}{d\gamma^1}, \quad b_2 = \frac{1}{2!}\frac{d^2 p_2(0)}{d\gamma^2}, \quad b_3 = \frac{1}{3!}\frac{d^3 p_2(0)}{d\gamma^3}, \quad b_4 = \frac{1}{4!}\frac{d^4 p_2(0)}{d\gamma^4}.$$

Now focusing attention on B_d, recall that $B_{O_2d} = A_{O_2}^{-1}(A_{O_2d} - I)B_{O_2}$ (see (4.6)). Since A_{O_2d} is already found, $A_{O_2}^{-1}$ is now computed. The inverse of A_{O_2} is given by

$$A_{O_2}^{-1} = T_D T_{O_2} \begin{bmatrix} -\frac{1}{T_{O_2}} & -\frac{1}{T_D} \\ 0 & -\frac{1}{T_D} \end{bmatrix} = \begin{bmatrix} -T_D & -T_{O_2} \\ 0 & -T_{O_2} \end{bmatrix} = \begin{bmatrix} -\frac{80}{N_e} & -T_{O_2} \\ 0 & -T_{O_2} \end{bmatrix}. \quad (4.14)$$

Thus, $A_{O_2}^{-1}$ can be represented with the following lower LFT:

$$A_{O_2}^{-1} = \mathscr{F}_\ell \left(M_{A_{O_2}^{-1}}, \frac{1}{N_e} \right), \quad (4.15)$$

where

$$M_{A_{O_2}^{-1}} = \begin{bmatrix} 0 & -T_{O_2} & -80 \\ 0 & -T_{O_2} & 0 \\ \hline 1 & 0 & 0 \end{bmatrix}. \quad (4.16)$$

To normalize $\frac{1}{N_e}$ to γ, the following upper LFT is used:

$$\frac{1}{N_e} = \mathscr{F}_u(M_\gamma, \gamma), \quad \text{where} \quad M_\gamma = \begin{bmatrix} 1 & 1 \\ \hline \frac{2}{N_{e,0}} & \frac{1}{N_{e,0}} \end{bmatrix}. \quad (4.17)$$

The approximated state–space matrices $\hat{A}_{O_2 d}$ and $\hat{B}_{O_2 d}$ are represented in Fig. 4.3 by their respective dotted boxes. The approximated state matrix $\hat{A}_{O_2 d}$ block is formed by the lower LFT $M_{A_{O_2 d}}$ connected to the time-varying parameter matrix $\gamma_k I_4$. The approximated input matrix $\hat{B}_{O_2 d}$ block is formed by the matrix multiplications of $B_{O_2 d}$ in (4.6). The $\hat{A}_{O_2 d}$, $\hat{B}_{O_2 d}$, and $C_{O_2 d}$ blocks are then connected in the standard state–space interconnection [54]. After performing the interconnection displayed in Fig. 4.3, the fourth-order approximated system used for controller design is given by

$$\hat{x}_{O_2}(k+1) = \hat{A}_{O_2 d}(\gamma_k)\hat{x}_{O_2}(k) + \hat{B}_{O_2 d}(\gamma_k)y(k),$$
$$\hat{y}_s(k) = C_{O_2 d}\hat{x}_{O_2}(k), \quad (4.18)$$

where

$$\hat{A}_{O_2 d}(\gamma_k) = \begin{bmatrix} \exp(-\frac{120}{d}) & a(\gamma_k) \\ 0 & b(\gamma_k) \end{bmatrix},$$

$$\hat{B}_{O_2 d}(\gamma_k) = \begin{bmatrix} \frac{d(\gamma_k+1)}{v_0(\gamma_k-1)} \left(\frac{a(\gamma_k)}{T_{O_2}} \right) - b(\gamma_k) + 1 \\ 1 - b(\gamma_k) \end{bmatrix}.$$

The approximated state matrix $\hat{A}_{O_2 d}(\gamma_k)$ follows directly from (4.8). The approximated input matrix $\hat{B}_{O_2 d}(\gamma_k)$ follows from the matrix operations performed to compute $B_{O_2 d}$ in (4.6). The following polynomial functions $a(\gamma_k)$ and $b(\gamma_k)$:

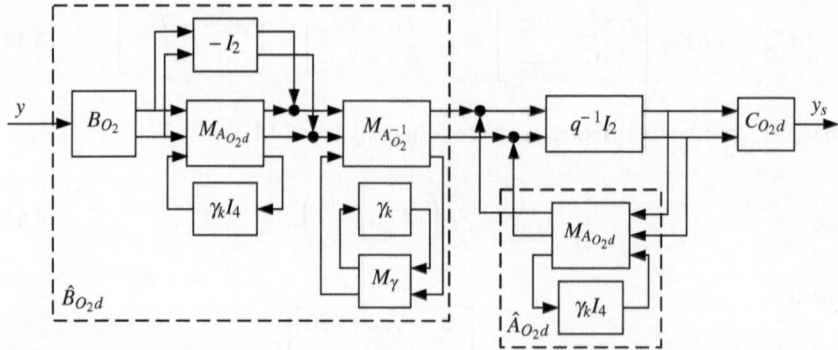

Fig. 4.3 Block diagram of the combined dynamics of the exhaust gas and sensor delays

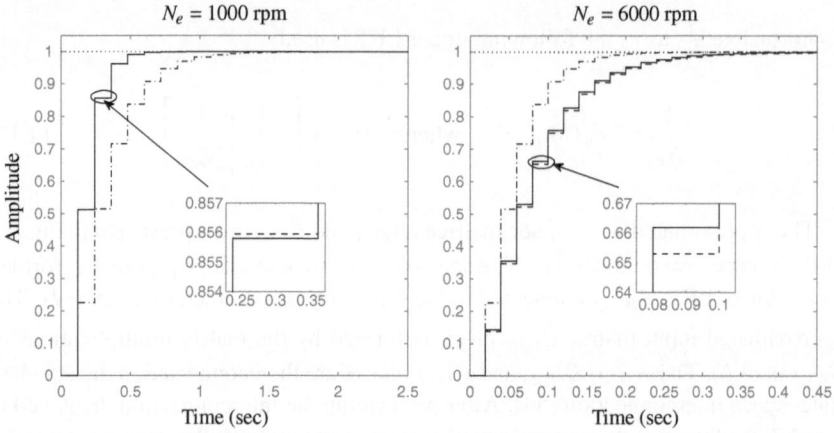

Fig. 4.4 Comparison of the response to a unit step function for the 4th order Taylor series approximation model in (4.18) (*dashed line*) and a model (4.6) with the engine speed fixed at 3,500 rpm (*dash-dot line*) to the exact discretized oxygen sensor delay model in (4.6) (*solid line*) at 1,000 and 6000 rpm

$$a(\gamma_k) = 0.3972 - 0.4891\gamma_k - 0.0984\gamma_k^2 + 0.0608\gamma_k^3 + 0.0975\gamma_k^4,$$
$$b(\gamma_k) = 0.3114 - 0.7266\gamma_k + 0.1211\gamma_k^2 + 0.3095\gamma_k^3 + 0.2231\gamma_k^4,$$

were found when selecting an oxygen sensor time constant of $T_{O_2} = 0.06\,$s and a transport delay of $T_D = \frac{80}{N_e}$, by setting $d = 80$, indicating that the transport delay is about 54 ms at an engine speed of 1,500 rpm. This was determined empirically through engine calibration tests.

To demonstrate the effectiveness of the proposed model for the event-based sampling of the oxygen sensor delay, a comparison is made between the proposed 4th order Taylor series approximation model and a fixed model computed at the nominal engine speed (3,500 rpm). In Fig. 4.4, the step response of the 4th order Taylor series approximation model (dashed line) is compared to the exact discretized model (solid line) at engine speeds of 1,000 and 6,000 rpm. The fixed model computed at the nominal engine speed (dash-dot line) is also compared to the exact model in Fig. 4.4. It is clear that the fixed model computed at the nominal engine speed either responds too slowly when the engine speed is less than the nominal speed or too quickly when the engine speed is greater than the nominal speed. However, the approximated model's response very closely follows the exact model's response as shown in Fig. 4.4.

4.2.4 An LPV System

In summary, by combining the wall-wetting dynamics in (4.2) and the oxygen sensor delay and dynamics in (4.18) as shown in Fig. 4.2, we obtain the following LPV system for the event-based discrete-time port-fuel-injection and oxygen sensor dynamics:

$$
\begin{bmatrix} x_{\mathrm{ww}}(k+1) \\ x_{\mathrm{comb}}(k+1) \\ \hat{x}_{O_2}(k+1) \end{bmatrix} = \begin{bmatrix} 1-\alpha_k & 0 & 0 \\ \frac{c\alpha_k}{m_A^0} & 0 & 0 \\ 0 & \hat{B}_{O_2 d}(\gamma_k) & \hat{A}_{O_2 d}(\gamma_k) \end{bmatrix} \begin{bmatrix} x_{\mathrm{ww}}(k) \\ x_{\mathrm{comb}}(k) \\ \hat{x}_{O_2}(k) \end{bmatrix}
$$

$$
+ \begin{bmatrix} 1-\beta_k \\ \frac{c\beta_k}{m_A^0} \\ 0 \end{bmatrix} m_i(k) + \begin{bmatrix} 0 \\ c \\ 0 \end{bmatrix} w_1(k), \qquad (4.19)
$$

$$
z(k) = \begin{bmatrix} 0 & 0 & -C_{O_2 d} \end{bmatrix} \begin{bmatrix} x_{\mathrm{ww}}(k) \\ x_{\mathrm{comb}}(k) \\ \hat{x}_{O_2}(k) \end{bmatrix} + w_2(k),
$$

where $x_{\mathrm{ww}}(k) = m_w(k-1)$ and $x_{\mathrm{comb}}(k)$ are the wall-wetting state and the combustion state for the system in the dotted box in Fig. 4.2.

As can be seen from (4.5), (4.6), (4.18), and (4.19), to apply the model of the LPV system, one needs to identify T_D and T_{O_2} (which are shown in Table 4.1) and measurable time-varying parameters such as α, β, and γ (which are shown in Table 4.2), which will be used for scheduling the gain of the controller. In particular, the identified bounds of scheduling variables ($[\underline{\alpha}, \overline{\alpha}]$, $[\underline{\beta}, \overline{\beta}]$ and $[\underline{\gamma}, \overline{\gamma}]$) as shown in Table 4.2 will be used in synthesizing the gain-scheduling controller. From now on, a compact notation Θ will denote an appropriate gain-scheduling matrix that contains the scheduling variables. The specific structure of Θ will be presented in (4.24) of Sect. 4.3.2. In addition, the LPV system in (4.19) is denoted by $P(\Theta)$. In the following section, we illustrate how to design the LPV gain-scheduling controller as a function of Θ for the LPV model developed in this section.

Table 4.1 Modeling parameters

Parameter	Value used in study
T_D is a function of engine speed, N_e	$80/N_e$
T_{O_2} is a constant	0.06

Table 4.2 Measurable time-varying parameters (scheduling parameters)

α(cylinder head temperature(t), manifold absolute pressure(t)) $\in [0.081, 0.1]$

β(cylinder head temperature(t), manifold absolute pressure(t)) $\in [0.28, 0.89]$

$\gamma(N_e(t)) = \dfrac{\frac{1}{N_e(t)} - \frac{1}{N_{e,0}}}{\frac{1}{N_e(t)} + \frac{1}{N_{e,0}}} \in [-0.55556, 0.26316]$

4.3 LPV Gain-Scheduling Controller Design

4.3.1 Control Strategy

The objective of the control system is to regulate the equivalence ratio y to a reference input w_2 using feed-forward and feedback control against the disturbance signal w_1 (See Fig. 4.2) and the time-varying wall-wetting dynamics. In particular, we want to guarantee the stability of the closed-loop system and also minimize the effect of the disturbances for any conceivable wall-wetting dynamics variations. The proposed control architecture is illustrated in Fig. 4.5. This scheme has five possible components, that is a feedback controller $K(\Theta)$, a feed-forward controller $K_f(\Theta)$, a filter $L(q)$, an integrator $I(q)$, and possibly a differentiator $D(q)$ (if a PID controller is desired).

The feedback controller $K(\Theta)$ will be designed for the generalized plant (solid box of Fig. 4.5), after selecting $K_f(\Theta)$, $L(q)$, $I(q)$, possibly $D(q)$, and weighting functions $W_1(q)$ and $W_2(q)$. Next, we will explain how to select these functions. After the selection, we will derive the generalized plant in Sect. 4.3.4 and we will synthesize $K(\Theta)$ for the derived generalized plant in Sect. 4.3.5.

The feed-forward controller $K_f(\Theta)$ is designed using the inverse of $cG(q)$

$$K_f(\Theta) = \frac{G^{-1}(q)}{c} = \frac{1}{c}\left(\frac{1 - (1 - \alpha_k)q^{-1}}{\beta_k + (\alpha_k - \beta_k)q^{-1}}\right).$$

The selection of the inverse of the plant as a feed-forward controller is a standard technique [54]. The input to the feed-forward controller is the mass of the air m_A, which can be measured online, multiplied by the equivalence ratio set point w_2. This is denoted by w_3, such that $w_3 = w_2 m_A$. $L(q)$ is designed as a low-pass filter such that the error output $z(k)$ is filtered with it

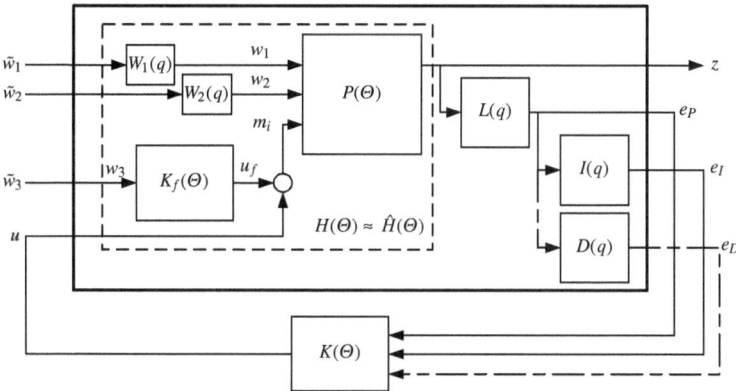

Fig. 4.5 The proposed control strategy for the fuel injection process (without the weighting functions $W_1(q)$ and $W_2(q)$). Weighting functions $W_1(q)$ and $W_2(q)$ are only used for controller synthesis. A first-order Taylor series expansion is applied to the systems inside the *dashed box* and the LPV control strategy is applied to all of the systems inside the *bold box*

$$L(q) = \frac{0.9999}{q - 0.0001405}.$$

The reason to filter the error output is that the control synthesis technique given by [11] requires that the output matrix be independent of the time-varying parameters and the measurement for control must not be corrupted by the unweighted exogenous input, $\tilde{w}(k)$ of the generalized plant. The low-pass filtering for this purpose is a standard procedure [2]. The low-pass filter $L(q)$ was obtained from the discretization of the following first-order continuous transfer function:

$$L^c(s) = \frac{2\pi f_c}{s + 2\pi f_c}$$

with a sample period of $\frac{120}{N_{e,0}}$. The cut-off frequency f_c of $L^c(s)$ was selected to be 20 Hz, which is high enough to obtain low error between the intended output of the continuous-time filter $L^c(q)$ and the observed output of the discrete-time filter $L(q)$ at different engine speeds, since the sampling rate is engine speed dependent. The filtered output is also integrated using the integrator

$$I(q) = \frac{1}{q - 1}$$

to obtain zero steady-state error. To enhance the response of the closed-loop system when large changes in w_1 are present, then derivative action [5]

$$D(q) := \frac{e_D(k)}{e_P(k)} = \frac{F(q - 1)}{(F + 1)q - 1}$$

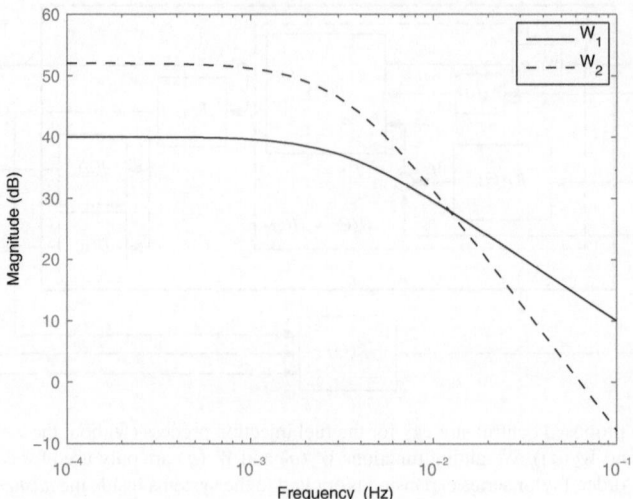

Fig. 4.6 Magnitude plot of the weighting functions W_1 and W_2

is introduced, where F is chosen to set the location of the pole of the derivative filter. Notice that $I(q)$ and $D(q)$ are not functions of the sampling rate, t_s. This is due to the requirement that, as previously stated, that the output matrix $\hat{C}_2$ be independent of the time-varying parameters. For this reason, $I(q)$ is really just a numerical summation and $D(q)$ is a filtered, numerical differencer.

To use ℓ_2 to ℓ_2 gain or $\mathcal{H}_\infty$ norm [72] for the performance criterion for shaping the frequency response of the closed-loop system, weighing functions (which can be considered design parameters) are also introduced in Fig. 4.5. The weighting functions are selected in the continuous-time domain as

$$W_1^c(s) = \frac{100}{50s + 1},$$

$$W_2^c(s) = \left(\frac{20}{50s + 1}\right)^2.$$

The bandwidth (or cut-off frequency) of each weighting function is very small and the DC gain is large, as shown in Fig. 4.6. The weighting functions are selected to model the frequency content of their respective input. For the fuel-to-air ratio disturbance w_1, the weighting function $W_1^c(s)$ is selected as a simple first-order low-pass filter to place an emphasis on low frequency disturbances, such as a step throttle change. The weighting function $W_2^c(s)$ is chosen to be a second-order low-pass filter with a high DC gain (4 times larger than that of $W_1^c(s)$) to provide more weight on the low frequency signals since w_2 is the step input of the desired equivalence ratio. To incorporate the weighting functions $W_1^c(s)$ and $W_2^c(s)$ into the discrete time system, they were discretized at a sample period of $\frac{120}{N_{e,0}}$ to obtain the following discrete-time weighting functions:

$$W_1(q) = \frac{0.1411}{q - 0.9986},$$

$$W_2(q) = \frac{0.0003982q + 0.0003979}{q^2 - 1.997178q + 0.997180}.$$

The input to each of the weighting functions are the unweighted exogenous inputs which are denoted by $\tilde{w}_1$, $\tilde{w}_2$, and $\tilde{w}_3$. Since there is no weighting function for w_3, in this case $\tilde{w}_3 = w_3$; which means that it is weighted equally at all frequencies. Notice that the weighting functions are chosen by the expected system inputs and their relative (frequency) importance, and they are only used for controller synthesis [54, 73].

4.3.2 Feed-Forward Compensated Generalized Plant

The feed-forward compensated generalized plant is denoted by $H(\Theta)$. As depicted in the dashed box of Fig. 4.5, the feed-forward compensated generalized plant consists of the feed-forward controller $K_f(\Theta)$, the plant $P(\Theta)$, and the weighting functions $W_1(q)$ and $W_2(q)$. The components of the feed-forward controller $K_f(\Theta)$ and the plant $P(\Theta)$ are illustrated in Fig. 4.7. The feed-forward controller $K_f(\Theta)$ components are encased inside the dashed box in Fig. 4.7 and the plant $P(\Theta)$ components are outside the dashed box.

In the feed-forward control compensated generalized plant $H(\Theta)$, the time-varying parameters α_k and β_k are equivalently transformed to a constant nominal value plus a time-varying fluctuation. For instance, the parameter variation of $\alpha_k \in [\underline{\alpha}, \overline{\alpha}]$ with $\alpha_0 = \frac{\underline{\alpha} + \overline{\alpha}}{2}$ is represented by

$$\alpha_\delta(k) = \alpha_k - \alpha_0 \in [\underline{\alpha} - \alpha_0, \overline{\alpha} - \alpha_0],$$

so that the parameter range of $\alpha_\delta(k)$ is centered around zero. Hence, α_k is replaced by $\alpha_0 + \alpha_\delta(k)$. The same is done for $\beta_k \in [\underline{\beta}, \overline{\beta}]$ as well. The parameter variation of N_e is represented by γ as shown in Eq. (4.10). The upper LFTs (see Appendix A) inside the dotted box in Fig. 4.7, $M_{\frac{1}{\beta}}$ and $M_{\frac{\alpha}{\beta}}$ are used to isolate the time-varying parameters $\beta_\delta(k)$ and $\alpha_\delta(k)$ [73]. β_δ is isolated from $\frac{1}{\beta_k}$ by

$$\frac{1}{\beta_k} = \frac{1}{\beta_0 + \beta_\delta(k)} = \mathscr{F}_u\left(M_{\frac{1}{\beta}}, \beta_\delta(k)\right), \tag{4.20}$$

where

$$M_{\frac{1}{\beta}} = \left[\begin{array}{c|c} -\frac{1}{\beta_0} & -\frac{1}{\beta_0} \\ \hline \frac{1}{\beta_0} & \frac{1}{\beta_0} \end{array}\right]. \tag{4.21}$$

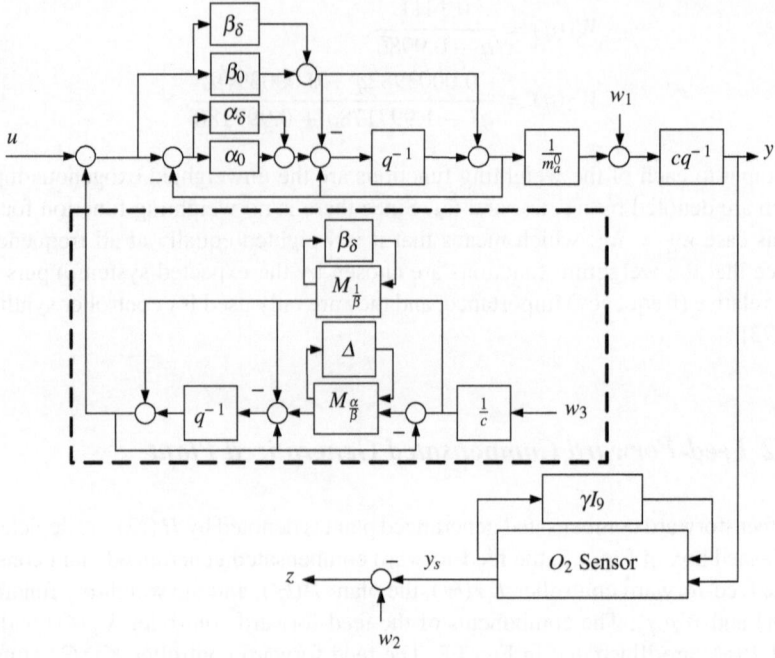

Fig. 4.7 The feed-forward control compensated generalized plant with the time-varying parameters included

Both $\beta_\delta(k)$ and $\alpha_\delta(k)$ are isolated from $\frac{\alpha_k}{\beta_k}$ by

$$\frac{\alpha_k}{\beta_k} = \frac{\alpha_0 + \alpha_\delta(k)}{\beta_0 + \beta_\delta(k)} = \mathscr{F}_u\left(M_{\frac{\alpha}{\beta}}, \Delta(k)\right),\qquad(4.22)$$

where

$$M_{\frac{\alpha}{\beta}} = \begin{bmatrix} -\frac{1}{\beta_0} & -\frac{1}{\beta_0} & -\frac{\alpha_0}{\beta_0} \\ 0 & 0 & 1 \\ \hline \frac{1}{\beta_0} & \frac{1}{\beta_0} & \frac{\alpha_0}{\beta_0} \end{bmatrix}, \quad \text{and} \quad \Delta(k) = \begin{bmatrix} \beta_\delta(k) & 0 \\ 0 & \alpha_\delta(k) \end{bmatrix}.$$

With the parameter variation represented in this way, the system is written as a discrete-time LPV system with LFT parameter dependency,

$$\begin{bmatrix} x(k+1) \\ l(k) \\ z(k) \end{bmatrix} = \begin{bmatrix} A & B_0 & B_1 & B_2 \\ C_0 & D_{00} & D_{01} & D_{02} \\ C_1 & D_{10} & D_{11} & D_{12} \end{bmatrix} \begin{bmatrix} x(k) \\ p(k) \\ \tilde{w}(k) \\ u(k) \end{bmatrix},$$

$$p(k) = \Theta(k)l(k),\qquad(4.23)$$

where $x(k) \in \mathbb{R}^n$ is the state at time k, $\tilde{w}(k) \in \mathbb{R}^r$ is the unweighted exogenous input, $z(k) \in \mathbb{R}^p$ is the error output, $p(k), l(k) \in \mathbb{R}^{n_p}$ are the pseudo-input and pseudo-output connected by $\Theta(k)$, and $u(k) \in \mathbb{R}^m$ is the control input. The state–space matrices for the LPV system in (4.23) are provided in Appendix B.

The time-varying parameter Θ in (4.23) follows the structure

$$\Theta \in \Theta = \{\operatorname{diag}(\beta_\delta I_3, \alpha_\delta I_2, \gamma I_9) : |\alpha_\delta| \le \delta_1, |\beta_\delta| \le \delta_2, |\gamma| \le 1\}, \qquad (4.24)$$

where $\delta_1 = \frac{\overline{\alpha} - \underline{\alpha}}{2}$ and $\delta_2 = \frac{\overline{\beta} - \underline{\beta}}{2}$.

4.3.3 First-Order Taylor Series Expansion of the LPV System

By inspection of the LPV system in (4.23), D_{00} was found to be a nonzero matrix. Hence, the system matrices are not affine functions, i.e., a linear combination of the time-varying parameters plus a constant translation. It is noted at this juncture that LPV control techniques exist which do handle rational parameter variation, namely the method developed by [65]. However, for discrete-time systems, no controller formula covering all parameter variation is given by [65]. Instead, for each set of parameters a controller must be solved for using the method given by [22]. Since a different controller is needed for each set of parameters, gridding over the parameter space [1] is necessary, which increases the complexity of implementing the controller in practice. In contrast, the method developed by [11] does not require any gridding over the parameter space. Also, as shown in (4.24) and Table 4.2 each of the parameters are less than 1 at all times. Therefore, neglecting the higher order parameter variation is a justifiable approximation. Hence, to utilize the control synthesis technique given by [11] as presented in Lemma 2.7, we calculate the first-order Taylor series approximation of the system matrices to obtain affine functions in Θ as demonstrated in Sect. 2.1.1. To find the Taylor series expansion, first the LFT (4.23) is rearranged to the following representation:

$$\begin{bmatrix} l(k) \\ x(k+1) \\ z(k) \end{bmatrix} = \underbrace{\begin{bmatrix} D_{00} & C_0 & D_{01} & D_{02} \\ B_0 & A & B_1 & B_2 \\ D_{10} & C_1 & D_{11} & D_{12} \end{bmatrix}}_{=:M} \begin{bmatrix} p(k) \\ x(k) \\ \tilde{w}(k) \\ u(k) \end{bmatrix}, \qquad (4.25)$$

$$p(k) = \Theta(k)l(k).$$

Notice that (4.25) is an upper LFT, i.e.,

$$H(\Theta) := \mathscr{F}_u(M, \Theta)$$

$$= \begin{bmatrix} A & B_1 & B_2 \\ C_1 & D_{11} & D_{12} \end{bmatrix} + \begin{bmatrix} B_0 \\ D_{10} \end{bmatrix} \Theta (I - D_{00}\Theta)^{-1} \begin{bmatrix} C_0 & D_{01} & D_{02} \end{bmatrix}. \qquad (4.26)$$

Using the Taylor series expansion at $\Theta = 0$, the system can be approximated as

$$\hat{H}(\Theta) = H(0) + \alpha_\delta \left[\nabla_{\alpha_\delta} H(0) \right] + \beta_\delta \left[\nabla_{\beta_\delta} H(0) \right] + \gamma \left[\nabla_\gamma H(0) \right],$$
$$=: \begin{bmatrix} \hat{A}(\alpha_\delta, \beta_\delta, \gamma) & \hat{B}_1(\alpha_\delta, \beta_\delta, \gamma) & \hat{B}_2(\alpha_\delta, \beta_\delta, \gamma) \\ \hat{C}_1(\alpha_\delta, \beta_\delta, \gamma) & \hat{D}_{11}(\alpha_\delta, \beta_\delta, \gamma) & \hat{D}_{12}(\alpha_\delta, \beta_\delta, \gamma) \end{bmatrix}, \tag{4.27}$$

where the relationship between α_δ, β_δ, and γ, and Θ can be found in (4.24) and $[\nabla_a H(0)]$ is the partial derivative of the LFT system $H(\Theta)$ in (4.26) with respect to a, which can be calculated as shown in Sect. 2.1.1. The state–space representation after performing the Taylor series expansion is given by

$$\begin{bmatrix} x(k+1) \\ z(k) \end{bmatrix} = \begin{bmatrix} \hat{A}(\alpha_\delta, \beta_\delta, \gamma) & \hat{B}_1(\alpha_\delta, \beta_\delta, \gamma) & \hat{B}_2(\alpha_\delta, \beta_\delta, \gamma) \\ \hat{C}_1(\alpha_\delta, \beta_\delta, \gamma) & \hat{D}_{11}(\alpha_\delta, \beta_\delta, \gamma) & \hat{D}_{12}(\alpha_\delta, \beta_\delta, \gamma) \end{bmatrix} \begin{bmatrix} x(k) \\ \tilde{w}(k) \\ u(k) \end{bmatrix}. \tag{4.28}$$

4.3.4 An Augmented LPV System for Synthesis

To create an appropriate measurement for gain-scheduling control, the LPV system $\hat{H}(\Theta)$ must be augmented with the low-pass filter $L(q)$, the integrator $I(q)$, and the numerical differencer $D(q)$ (when designing a gain-scheduled PID controller).

4.3.4.1 PI Control

After augmenting the affine LPV system with the low pass filter and the integrator the augmented state–space representation is given by

$$\begin{bmatrix} x_{\text{AUG}}(k+1) \\ z(k) \\ e(k) \end{bmatrix} = \begin{bmatrix} \tilde{A}(\alpha_\delta, \beta_\delta, \gamma) & \tilde{B}_1(\alpha_\delta, \beta_\delta, \gamma) & \tilde{B}_2(\alpha_\delta, \beta_\delta, \gamma) \\ \tilde{C}_1(\alpha_\delta, \beta_\delta, \gamma) & \tilde{D}_{11}(\alpha_\delta, \beta_\delta, \gamma) & \tilde{D}_{12}(\alpha_\delta, \beta_\delta, \gamma) \\ \tilde{C}_2 & 0 & 0 \end{bmatrix} \begin{bmatrix} x_{\text{AUG}}(k) \\ \tilde{w}(k) \\ u(k) \end{bmatrix} \tag{4.29}$$

where the augmented states are given by $x_{\text{AUG}}(k) = \left[x(k)^\mathsf{T} \, x_L(k) \, x_I(k) \right]^\mathsf{T} \in \mathbb{R}^{n_{\text{AUG}}}$ with $n_{\text{AUG}} = n + 2$, and the measurement for control is given by $e(k) = [e_P(k) \, e_I(k)]^\mathsf{T} \in \mathbb{R}^q$ with $q = 2$. The state–space matrices are given by

$$\tilde{A}(\alpha_\delta, \beta_\delta, \gamma) = \begin{bmatrix} \hat{A}(\alpha_\delta, \beta_\delta, \gamma) & 0 & 0 \\ B_L \hat{C}_1(\alpha_\delta, \beta_\delta, \gamma) & A_L & 0 \\ 0 & & C_L & 1 \end{bmatrix},$$

$$\tilde{B}_1(\alpha_\delta, \beta_\delta, \gamma) = \begin{bmatrix} \hat{B}_1(\alpha_\delta, \beta_\delta, \gamma) \\ B_L \hat{D}_{11}(\alpha_\delta, \beta_\delta, \gamma) \\ 0 \end{bmatrix},$$

$$\tilde{B}_2(\alpha_\delta, \beta_\delta, \gamma) = \begin{bmatrix} \hat{B}_2(\alpha_\delta, \beta_\delta, \gamma) \\ B_L \hat{D}_{12}(\alpha_\delta, \beta_\delta, \gamma) \\ 0 \end{bmatrix},$$

$$\tilde{C}_1(\alpha_\delta, \beta_\delta, \gamma) = \begin{bmatrix} \hat{C}_1(\alpha_\delta, \beta_\delta, \gamma) \, 0 \, 0 \end{bmatrix},$$

$$\tilde{C}_2 = \begin{bmatrix} 0 \, C_L \, 0 \\ 0 \, 0 \, 1 \end{bmatrix},$$

and $\tilde{D}_{11}(\alpha_\delta, \beta_\delta, \gamma) = \hat{D}_{11}(\alpha_\delta, \beta_\delta, \gamma)$, $\tilde{D}_{12}(\alpha_\delta, \beta_\delta, \gamma) = \hat{D}_{12}(\alpha_\delta, \beta_\delta, \gamma)$. The matrices (A_L, B_L, C_L) represent the state-space matrices of the low-pass filter $L(q)$.

4.3.4.2 PID Control

When designing a gain-scheduling PID controller, the augmented affine LPV system with the low pass filter, the integrator, and the numerical differencer, the augmented state space representation is given by

$$\begin{bmatrix} x_{AUG}(k+1) \\ z(k) \\ e(k) \end{bmatrix} = \begin{bmatrix} \tilde{A}(\alpha_\delta, \beta_\delta, \gamma) & \tilde{B}_1(\alpha_\delta, \beta_\delta, \gamma) & \tilde{B}_2(\alpha_\delta, \beta_\delta, \gamma) \\ \tilde{C}_1(\alpha_\delta, \beta_\delta, \gamma) & \tilde{D}_{11}(\alpha_\delta, \beta_\delta, \gamma) & \tilde{D}_{12}(\alpha_\delta, \beta_\delta, \gamma) \\ \tilde{C}_2 & 0 & 0 \end{bmatrix} \begin{bmatrix} x_{AUG}(k) \\ \tilde{w}(k) \\ u(k) \end{bmatrix}$$

$$(4.30)$$

where the augmented states are given by $x_{AUG}(k) = \begin{bmatrix} x(k)^\mathsf{T} \, x_L(k) \, x_I(k) \, x_D(k) \end{bmatrix}^\mathsf{T} \in \mathbb{R}^{n_{AUG}}$ with $n_{AUG} = n + 3$, and the measurement for control is given by $e(k) = [e_P(k) \, e_I(k) \, e_D(k)]^\mathsf{T} \in \mathbb{R}^q$ with $q = 3$. The state–space matrices are given by

$$\tilde{A}(\alpha_\delta, \beta_\delta, \gamma) = \begin{bmatrix} \hat{A}(\alpha_\delta, \beta_\delta, \gamma) & 0 & 0 & 0 \\ B_L \hat{C}_1(\alpha_\delta, \beta_\delta, \gamma) & A_L & 0 & 0 \\ 0 & C_L & 1 & 0 \\ 0 & B_D C_L & 0 & A_D \end{bmatrix},$$

$$\tilde{B}_1(\alpha_\delta, \beta_\delta, \gamma) = \begin{bmatrix} \hat{B}_1(\alpha_\delta, \beta_\delta, \gamma) \\ B_L \hat{D}_{11}(\alpha_\delta, \beta_\delta, \gamma) \\ 0 \\ 0 \end{bmatrix},$$

$$\tilde{B}_2(\alpha_\delta, \beta_\delta, \gamma) = \begin{bmatrix} \hat{B}_2(\alpha_\delta, \beta_\delta, \gamma) \\ B_L \hat{D}_{12}(\alpha_\delta, \beta_\delta, \gamma) \\ 0 \\ 0 \end{bmatrix},$$

$$\tilde{C}_1(\alpha_\delta, \beta_\delta, \gamma) = \begin{bmatrix} \hat{C}_1(\alpha_\delta, \beta_\delta, \gamma) \, 0 \, 0 \, 0 \end{bmatrix},$$

$$\tilde{C}_2 = \begin{bmatrix} 0 & C_L & 0 & 0 \\ 0 & 0 & 1 & 0 \\ 0 & D_D C_L & 0 & C_D \end{bmatrix},$$

and $\tilde{D}_{11}(\alpha_\delta, \beta_\delta, \gamma) = \hat{D}_{11}(\alpha_\delta, \beta_\delta, \gamma)$, $\tilde{D}_{12}(\alpha_\delta, \beta_\delta, \gamma) = \hat{D}_{12}(\alpha_\delta, \beta_\delta, \gamma)$. The matrices (A_L, B_L, C_L) represent the state–space matrices of the low-pass filter $L(q)$ and the matrices (A_D, B_D, C_D, D_D) represent the state-space matrices of the filtered, numerical differencer $D(q)$.

4.3.5 A Gain-Scheduling Control Synthesis Problem

Having augmented all components for the controller synthesis, we need to synthesize the $\mathcal{H}_\infty$ gain-scheduling controller $K(\Theta)$. The ℓ_2 gain of the LPV system in (4.29) with a gain-scheduling feedback controller is defined as

$$\max_{\Theta \in \Theta, \|\tilde{w}\|_{\ell_2} \neq 0} \frac{\|z\|_{\ell_2}}{\|\tilde{w}\|_{\ell_2}}. \tag{4.31}$$

Now we formally state the gain-scheduling control design problem.

Problem   The goal is to design a static gain-scheduling control $u(k) = K(\Theta)e(k)$ that stabilizes the closed-loop system and minimizes the worst-case ℓ_2 gain ($\mathcal{H}_\infty$ norm) of the closed-loop LPV system in (4.31) for any trajectories of $\Theta(k) \in \Theta$.

The gain-scheduling method provided by [11] guarantees an $\mathcal{H}_\infty$ cost such that for any exogenous input $\tilde{w}$, the performance output z satisfies

$$\|z\|_{\ell_2} < \eta \, \|\tilde{w}\|_{\ell_2},$$

for any trajectories of $\Theta(k) \in \Theta$. This method was derived for discrete-time polytopic time-varying systems. Therefore, in the next section, we will transform the augmented, affine system into a polytopic time-varying system to synthesize the controller.

4.3.6 Controller Synthesis for Polytopic Linear Time-Varying System

The augmented state–space representation $(\tilde{A}(\alpha_\delta, \beta_\delta, \gamma), \tilde{B}_1(\alpha_\delta, \beta_\delta, \gamma), ...)$ in either (4.29) or (4.30) can be converted into a discrete-time polytopic time-varying system $(\bar{A}[\lambda(k)], \bar{B}_1[\lambda(k)], ...)$ by using the state–space matrices at vertices $\{\mathcal{V}_i\}$ of the parameter space polytope displayed in Fig. 4.8. Any system inside of the convex

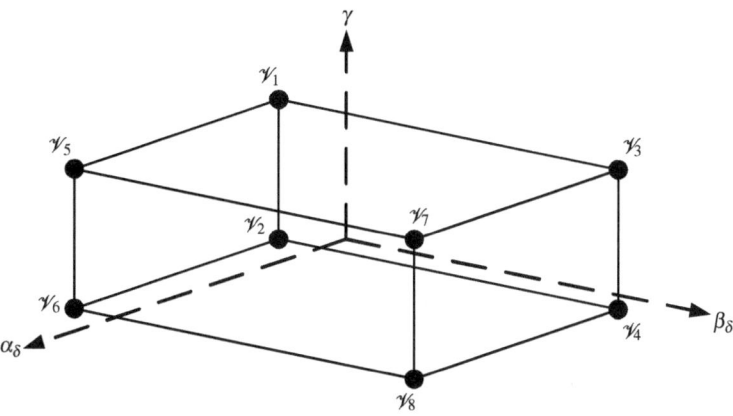

Fig. 4.8 Parameter space polytope

parameter set is represented by a convex combination of the vertex systems as weighted by the vector $\lambda(k)$ of barycentric coordinates. Barycentric coordinates are used to specify the location of a point as the center of mass, or barycenter, of masses placed at the vertices of a simplex. Warren et al. [60] provides a formula, which is covered in Sect. 2.1.2, for computing the barycentric coordinates for any convex polytope. The discrete-time polytopic LTV system is given by

$$\begin{bmatrix} x(k+1) \\ z(k) \\ e(k) \end{bmatrix} = \begin{bmatrix} \bar{A}[\lambda(k)] & \bar{B}_1[\lambda(k)] & \bar{B}_2[\lambda(k)] \\ \bar{C}_1[\lambda(k)] & \bar{D}_{11}[\lambda(k)] & \bar{D}_{12}[\lambda(k)] \\ \bar{C}_2 & 0 & 0 \end{bmatrix} \begin{bmatrix} x(k) \\ w(k) \\ u(k) \end{bmatrix},$$
$$e(k) = \begin{bmatrix} e_P(k) \; e_I(k) \end{bmatrix}^\mathsf{T}, \tag{4.32}$$

where, for all $k \in \mathbb{Z}_{\geq 0}$, $\lambda(k)$ is the vector of time-varying barycentric coordinates that belong to the unit simplex (2.12). A way to compute the barycentric coordinate vector $\lambda(k)$ for a given $\alpha_\delta(k)$, $\beta_\delta(k)$, and $\gamma(k)$ is provided in Sect. 2.1.2. For all $k \in \mathbb{Z}_{\geq 0}$, the rate of variation of the weights

$$\Delta\lambda_i(k) = \lambda_i(k+1) - \lambda_i(k), \quad i = 1, \ldots, N$$

is limited by the calculated bound $b \in [0, 1]$ such that the rate limit (2.27), reproduced here

$$-b\lambda_i(k) \leq \Delta\lambda_i(k) \leq b(1 - \lambda_i(k)), \quad i = 1, \ldots, N$$

is in effect.

The system matrices $\bar{A}[\lambda(k)] \in \mathbb{R}^{n \times n}$, $\bar{B}_1[\lambda(k)] \in \mathbb{R}^{n \times r}$, $\bar{B}_2[\lambda(k)] \in \mathbb{R}^{n \times m}$, $\bar{C}_1[\lambda(k)] \in \mathbb{R}^{p \times n}$, $\bar{D}_{11}[\lambda(k)] \in \mathbb{R}^{p \times r}$, $\bar{D}_{12}[\lambda(k)] \in \mathbb{R}^{p \times m}$ belong to the polytope

$$\mathfrak{D} = \{(\bar{A}, \bar{B}_1, \bar{B}_2, \bar{C}_1, \bar{D}_{11}, \bar{D}_{12})(\lambda(k)) : (\bar{A}, \bar{B}_1, \bar{B}_2, \bar{C}_1, \bar{D}_{11}, \bar{D}_{12})(\lambda(k))$$

$$= \sum_{i=1}^{N} \lambda_i(k)(\bar{A}, \bar{B}_1, \bar{B}_2, \bar{C}_1, \bar{D}_{11}, \bar{D}_{12})_i, \lambda(k) \in \Lambda_N\}.$$

The system matrices at any time k are the weighted summation of vertex system matrices $\{\mathscr{V}_i\}$ weighted by their barycentric coordinates $\lambda_i(k)$, i.e.

$$\bar{A}(k) = \sum_{i=1}^{N} \lambda_i(k)\bar{A}(\mathscr{V}_i), \quad i = 1, \ldots, N.$$

The same computation holds for $\bar{B}_1, \bar{B}_2, \bar{C}_1, \bar{D}_{11}$, and $\bar{D}_{12}$ as well.

Lemma 2.7 provides a finite set of LMIs that can be used to design the gain-scheduling controller. Due to Lemma 2.7, if there exists matrices $G_{i,1} \in \mathbb{R}^{q \times q}$, $G_{i,2} \in \mathbb{R}^{(n_{\mathrm{AUG}} - q) \times q}$, $G_{i,3} \in \mathbb{R}^{(n_{\mathrm{AUG}} - q) \times (n_{\mathrm{AUG}} - q)}$, $Z_{i,1} \in \mathbb{R}^{m \times q}$ and symmetric matrices $P_i \in \mathbb{R}^{n_{\mathrm{AUG}} \times n_{\mathrm{AUG}}}$ such that the LMI conditions in (2.51) and (2.52) are satisfied, the gain-scheduling static feedback control is then obtained as shown in (2.58). The LMIs in (2.51) and (2.52) are solved by programming them into MATLAB using the LMI lab solver [23], which is included in the Robust Control toolbox. The matrices $G_{i,1}, G_{i,2}, G_{i,3}, Z_{i,1}, P_i$, and the $\mathscr{H}_\infty$ cost η are programmed in MATLAB as free matrix variables for the LMI optimization to choose. During the solution process, the $\mathscr{H}_\infty$ cost η is minimized until the optimal solution is obtained.

4.4 Design of LTI Feedback Controller

The open-loop state–space plant used for designing this controller is the same as that in Fig. 4.7, but has the low-pass filter $L(q)$ and the integrator $I(q)$ added without performing any Taylor series expansion. Using the nominal parameters, the closed-loop state–space representation is

$$\begin{aligned} x(k+1) &= \mathscr{A}(K)x(k) + B_1 w(k), \\ z(k) &= \mathscr{C}(K)x(k) + D_{11} w(k), \end{aligned} \qquad (4.33)$$

where

$$\mathscr{A}(K) = A + B_2 K C_2 \quad \text{and} \quad \mathscr{C}(K) = C_1 + D_{12} K C_2.$$

Denoting the transfer function from w to z by H_{wz}, the inequality $\|H_{wz}\|_\infty^2 < \mu$ holds if, and only if, there exists a symmetric matrix P such that

$$
\begin{bmatrix}
P & \mathscr{A}(K)P & B_1 & 0 \\
P\mathscr{A}^T(K) & P & 0 & P\mathscr{C}^T(K) \\
B_1^T & 0 & I & D_{11}^T \\
0 & \mathscr{C}(K)P & D_{11} & \mu I
\end{bmatrix} \succ 0
\tag{4.34}
$$

is feasible [18]. The optimal feedback controller K for the closed-loop system (4.33) is formulated as the optimization of the bilinear matrix inequality (BMI)

$$
\min_{\mu,P,K} \; \mu \quad \text{subject to (4.34)}
\tag{4.35}
$$

where $P = P^T \in \mathbb{R}^{n \times n}$ and $K \in \mathbb{R}^{1 \times 2}$ for a PI controller or $K \in \mathbb{R}^{1 \times 3}$ for a PID controller. The BMI (4.35) was solved using the PENBMI software [34] as a MATLAB function in conjunction with the YALMIP [35] programming interface to find the fixed $\mathscr{H}_\infty$ PI controller $K_{PI} = [1.8260 \quad 0.3205]$ and the PID controller $K_{PID} = [1.4871 \quad 0.5009 \quad 0.8942]$.

4.5 Simulation Results

To validate the effectiveness of the proposed gain-scheduling controller, simulations are shown using the original plant in (4.23) for the following cases: engine cold start, load change, and engine speed change.

The benefit of a gain-scheduled controller is demonstrated by comparing its performance with that of a fixed gain $\mathscr{H}_\infty$ controller, which was designed for the nominal parameters as shown in Sect. 4.4.

In each simulation, the time-varying parameters α and β are corrupted by low-pass filtered white noise of up to 10 % their nominal values to represent the slowly drifting offset that might occur in practical situations. To see transient responses, the initial conditions of the plant for Case 1 were chosen such that a little extra fuel is injected at first, giving a slightly higher equivalence ratio than 1. The initial conditions in Cases 2 and 3 were set such that the plant would start with an equivalence ratio of 1. For the following simulation cases, the extracted profiles of time-varying parameters from engine dynamometer tests were used.

4.5.1 Case 1: Engine Cold Start

We simulate an engine operation when it was started with coolant temperature of $0\,^\circ\text{C}$ and heated to its normal operational coolant temperature of approximately $100\,^\circ\text{C}$ within about 2 min at an engine speed of 1,500 rpm. The purpose of this simulation is to emulate the cold start of an internal combustion engine when the engine is operated at high idle speed during the warm-up. Note that during the engine warm-up process the fuel vapor is much less at low temperature than that at high temperature. Therefore, this leads to quite different wall-wetting dynamics. The wall-wetting dynamics coefficients α and β defined in (4.3) were obtained from actual engine test data and they are functions of engine coolant temperature, speed, and load. Since speed and load are fixed in this simulation, both α and β are functions of engine temperature and their values are shown in Fig. 4.9e. Notice that the transient response at 25 s in Fig. 4.9 is due to the change in the wall-wetting parameters as shown in Fig. 4.9e. When the engine has been warming up for about 90 s, the closed-loop system with the fixed $\mathcal{H}_\infty$ controller becomes unstable, while the LPV controller remains stable. Thus, in Fig. 4.9a, one can readily see the LPV controller's advantage of guaranteed stability as the parameters vary with time.

4.5.2 Case 2: Load Change

In this case we simulate an engine dynamometer experiment for an engine operated at a temperature of 80 °C with an engine speed of 1,500 rpm. After the engine is stably operated at this condition with a 32 % throttle, the load is increased by a step throttle position from 32 to 46 %. Note that in the dynamometer test, the engine speed was maintained by a dynamometer control system by increasing the load torque. This is similar to the driving condition that a step throttle is applied to maintain the vehicle speed when the vehicle is driven up a hill. Note that the step increment of throttle position produces a slight change in the wall-wetting parameter β as shown in Fig. 4.10e. But in Fig. 4.10, one can find the benefit of guaranteed performance of the gain-scheduling controller over the time-varying parameters. Note that the step throttle occurred at the 30th second results in a momentary spike in the equivalence ratio due to the step air mass flow; but it is quickly pulled back into its target level by the gain-scheduled controller, while the fixed $\mathcal{H}_\infty$ controller takes much longer time with a lot of oscillations and uses more control effort.

4.5.3 Case 3: Engine Speed Change

In this simulation, an engine was operated in a dynamometer with its coolant temperature at 80 °C. To demonstrate the capability for the gain-scheduling controller to handle fast engine speed variations, smoothed step commands were applied to the engine dynamometer to manipulate the engine speeds shown in Fig. 4.11f. The resulting engine wall-wetting dynamic parameters, shown in Fig. 4.11e, were used in the simulation. In Fig. 4.11a, one can see that both controllers, gain-scheduling, and fixed $\mathcal{H}_\infty$, regulate the engine equivalence ratio to its target value of one within 5 % error except at 25th second when the engine speed was increased abruptly from 1,000 to 4,500 rpm. In this case, the engine equivalence ratio response converges to its target value smoothly for the gain-scheduling controller but with a lot of oscillations

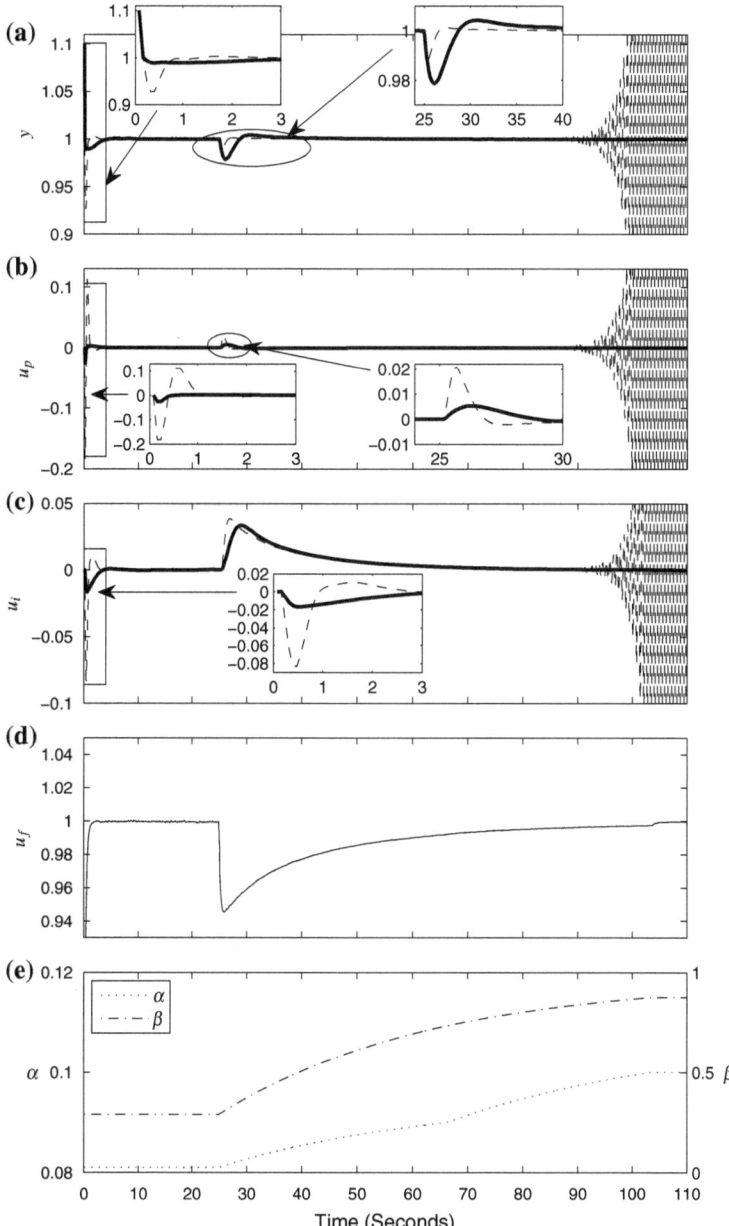

Fig. 4.9 Case 1 engine cold start: In plots **a, b, c**, and **d** the equivalence ratio $y(k)$, proportional control $u_p(k)$, integral control $u_i(k)$, and the feed-forward control are compared for the gain-scheduling feedback controller (*solid line*) and the fixed $\mathcal{H}_\infty$ controller (*dashed line*). The time-varying parameters α (*dotted line*, *left* axis) and β (*dash-dot line*, *right* axis) are displayed in plot **e**

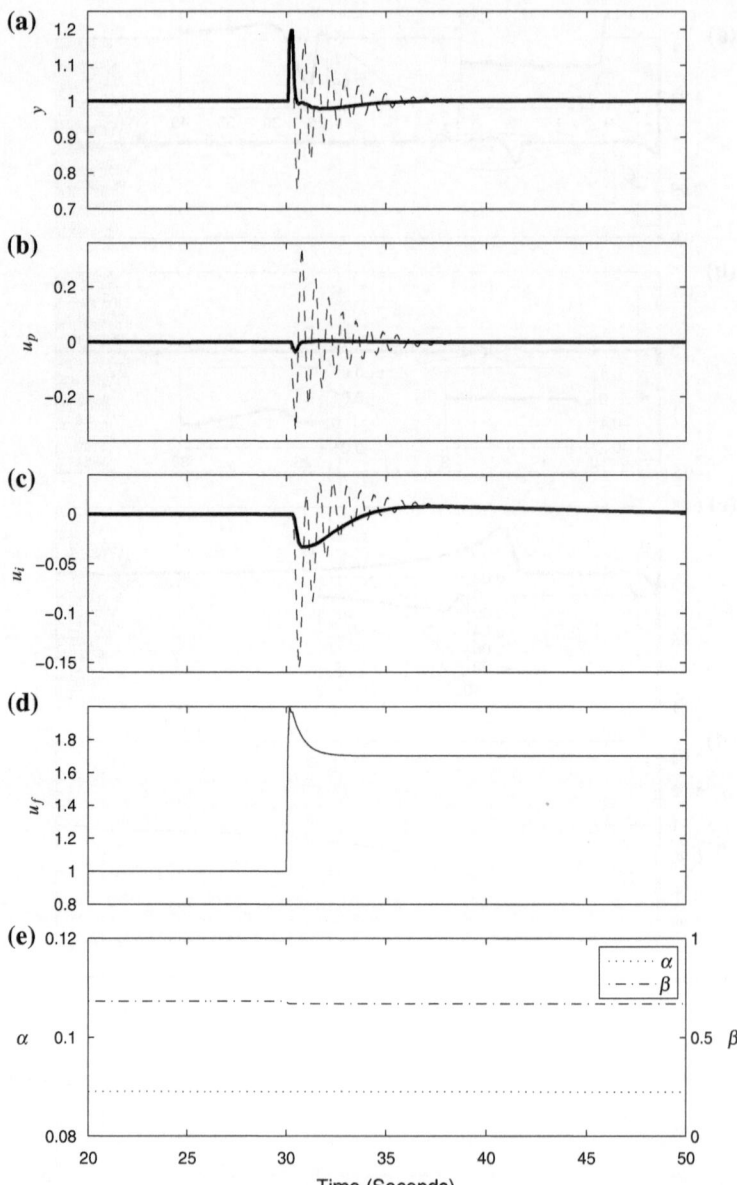

Fig. 4.10 Case 2 load change: In plots **a**, **b**, **c**, and **d** the equivalence ratio $y(k)$, proportional control $u_p(k)$, integral control $u_i(k)$, and the feed-forward control are compared for the gain-scheduling feedback controller (*solid line*) and the fixed $\mathcal{H}_\infty$ controller (*dashed line*). The time-varying parameters α (*dotted line*, *left axis*) and β (*dash-dot line*, *right axis*) are displayed in plot **e**

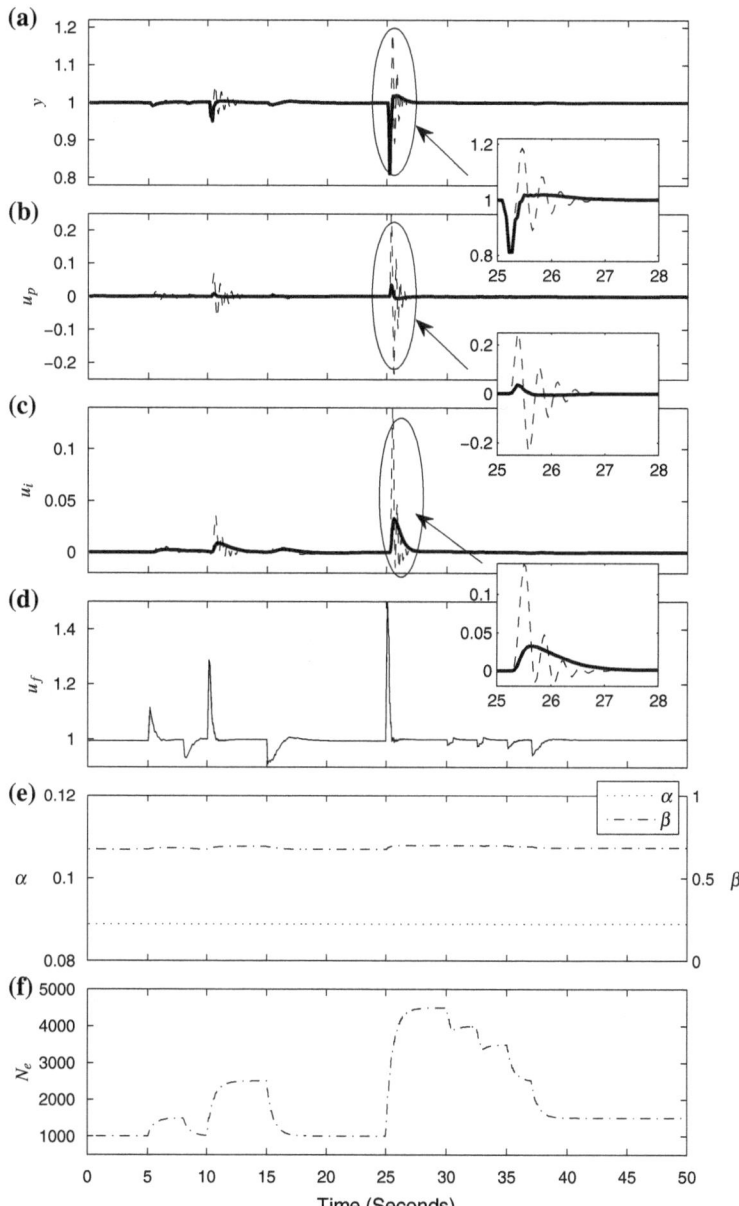

Fig. 4.11 Case 3 engine speed change: In plots **a, b, c,** and **d** the equivalence ratio $y(k)$, proportional control $u_p(k)$, integral control $u_i(k)$, and the feed-forward control are compared for the gain-scheduling feedback controller (*solid line*) and the fixed $\mathcal{H}_\infty$ controller (*dashed line*). The wall-wetting parameters α (*dotted line*, *left* axis) and β (*dash-dot line*, *right* axis) are displayed in plot **e**. The engine speed v is displayed in plot **f**

for the fixed $\mathcal{H}_\infty$ controller. This situation is similar to a transmission gear shifting where a rapid engine speed change may occur. Again, one can see the advantage of guaranteed performance over the time-varying parameters as the gain-scheduled controller regulates the equivalence ratio back into safe limits quicker and with less overshoot than the fixed $\mathcal{H}_\infty$ controller.

4.6 HIL Simulation

The engine model used for the HIL simulation is a control-oriented four-cylinder dual fuel mean-value engine model developed at Michigan State University [68], which satisfies the requirements of validating an engine controller. The term "mean-value" indicates that the developed engine model neglects the reciprocating behavior of the engine, assuming all processes and effects are spread out over the engine cycle. For the HIL simulation, this model describes the input–output behavior of the physical engine systems with reasonable simulation accuracy using relatively low computational throughput. Most of the dynamic equations used in our modeling work are from the reference book [29], which provides a good overview of engine modeling. This engine model also includes all engine transient dynamics. Figure 4.12 shows the overall mean-valve engine model architecture, along with main subsystem models, such as air-to-fuel ratio, manifold air pressure (MAP), brake mean effective pressure (BMEP), engine torque, exhaust temperature, etc.

4.6.1 Mean-Value Engine Models

The subsystems that are described mathematically by their averaged dynamic behaviors are given below.

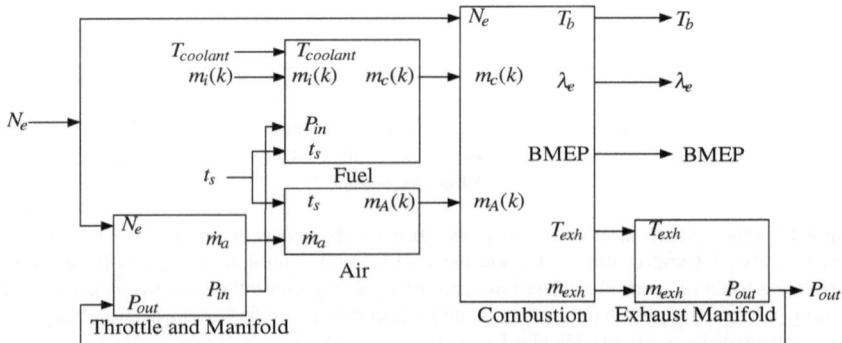

Fig. 4.12 Mean-value engine model

4.6.1.1 Valve Model

The valve model is used to compute the mass flow rate of air across the valve. The model used for the intake throttle and the EGR valve follow the governing equations

$$m_v = C_d(\theta)A(\theta)\frac{P_u}{\sqrt{RT_u}}\Psi\left(\frac{P_d}{P_u}\right),$$ (4.36)

and

$$\Psi\left(\frac{P_d}{P_u}\right) = \begin{cases} \sqrt{2\frac{P_d}{P_u}\left(1 - \frac{P_d}{P_u}\right)} & \text{if } \frac{1}{2} < \frac{P_d}{P_u} < 1, \\ \frac{1}{\sqrt{2}} & \text{if } \frac{P_d}{P_u} < \frac{1}{2}, \end{cases}$$ (4.37)

where C_d is the valve discharge coefficient; θ is the valve opening angle; R is the gas constant; A is the valve open area; P_u and T_u are the pressure and temperature upstream from the valve; and m_v is the mass flow rate across the valve. The governing equations (4.36), (4.37) follow the assumption that the spacial effects of the connecting pipes before and after the valve are neglected and that the thermodynamic characteristics of the connecting pipes are isentropic expansion.

4.6.1.2 Manifold Filling Dynamic Model

The manifold pressure of the intake and the exhaust is computed as a function of time by the governing equation

$$P_m(t) = P_m(0) + \int_0^t \frac{RT_m}{V_m}(m_{\text{in}} - m_{\text{out}})dt$$ (4.38)

where P_m is the manifold pressure; T_m is the manifold temperature; V_m is the manifold volume; m_{in} and m_{out} are the inlet and outlet air mass flow rates; and R is the universal gas constant. The assumptions made by the governing equation (4.38) are that the receiving behavior is an adiabatic process; the thermodynamic states are uniform over the manifold volume; and the manifold temperature is averaged over one engine cycle.

4.6.1.3 Engine Respiration Model

The mass flow rate of the air across of the engine cylinders, m_e, is computed by the engine respiration model

$$\dot{m}_e = \frac{P_{in}}{RT_{in}} \frac{V_d N_e}{30} \kappa \left(\frac{P_{out}}{P_{in}}, v \right) \tag{4.39}$$

where κ is a two degree of freedom look-up table; P_{in} and T_{in} are the mean pressure and temperature at the intake manifold; P_{out} is the mean pressure at the exhaust manifold; V_d is the engine displacement.

4.6.1.4 Crankshaft Dynamic Model

The crankshaft dynamic model, based on Newton's theory assuming a rigid crankshaft, is derived as

$$\dot{N}_e = \frac{60}{2\pi} \frac{T_b - T_l}{J_e} \tag{4.40}$$

where J_e is the rotational inertia of the engine crankshaft; and T_b and T_l are the engine brake and load torques. The desired engine speed is maintained by an engine dynamometer model that generates the engine load torque, T_l, using a feedback PID controller.

4.6.2 Event-Based Engine Models

The mathematical models used to simulate the cycle-to-cycle varying variables of engine subsystems are given below. Each variable in this section is updated based on the engine cycle (k) and is independent of time, t.

4.6.2.1 Event-Based Wall-Wetting Dynamics

When port-fuel-injection is used to deliver fuel to the engine cylinders, some of the fuel injected after each injector pulse enters the cylinders. However, the remaining fuel sticks to the walls of the intake port and on the back of the intake valve. The total fuel entering the engine cylinders then consists of fuel injected from the current injection pulse and fuel vapor from the fuel mass stored on the walls from previous injection pulses. Knowledge of this process is necessary to control the fuel metering for precise air-to-fuel ratio control. The event based wall-wetting dynamics used in the engine for HIL simulation are the same as those in (4.2).

4.6.2.2 Event-Based Engine Air-to-Fuel Ratio

The gas exchange behavior of the engine introduces dynamics into the air–fuel ratio calculation. Since the engine uses exhaust gas recirculation, a substantial amount of the burned gas remains in the cylinder. The gas fraction carries the air-to-fuel ratio of the previous engine cycle into the current cycle. Due to this behavior, the air–fuel ratio is modeled cycle-to-cycle as

$$\lambda_e(k) = \frac{\tilde{\lambda}_e(k)M_{\text{fresh}}(k) + \lambda_e(k-1)M_{\text{burnt}}(k)}{M_{\text{fresh}}(k) + M_{\text{burnt}}(k)}, \qquad (4.41)$$

where $\tilde{\lambda}_e$ is the normalized air-to-fuel ratio defined as

$$\tilde{\lambda}_e(k) = \frac{m_A(k)}{m_c(k)}\frac{1}{c}. \qquad (4.42)$$

λ_e is the normalized air-to-fuel ratio of the gas mixture inside the engine cylinder after the intake valve is closed. M_{fresh} is the mass of the fresh gas mixture charge in the cylinder, which is the summation of the fresh air mass m_A and the fresh fuel mass m_c, and M_{burnt} is the burned gas remaining in the engine cylinder after the exhaust valve closes, which includes burned gas due to both internal and external EGR (exhaust gas recirculation). Note that these dynamics are quite different from the LPV design model described in Fig. 4.2.

4.6.2.3 Event-Based Engine Brake Torque

Every combustion event, the engine brake torque calculation is triggered using the following equation:

$$T_b(k) = \frac{m_c(k)H_l n}{4\pi}\eta_e(N_e, \chi, \theta_{\text{st}}, x_{\text{EGR}}) \qquad (4.43)$$

where n is the number of engine cylinders; H_l is the lower heat value of the fuel; η_e is the engine efficiency, which is a function of engine speed, normalized air-to-fuel ratio, spark timing θ_{st}, and the exhaust-gas-recirculation rate x_{EGR}.

4.6.3 Setup and Implementation

The mean-value engine model was implemented into an Opal-RT HIL system using MATLAB/Simulink. The engine model was updated at a sample period of 1 ms. Similarly, the LPV controller, along with feedforward controller, was implemented as an event-based discrete controller in Simulink into a Mototron Engine Control Unit (ECU) sampled every 5 ms as a function call, see HIL simulation scheme shown in Fig. 4.13. The Opal-RT HIL simulator communicates with the Mototron ECU controller through the high speed controller-area-network (CAN), where signals were sent and received with minimal delay.

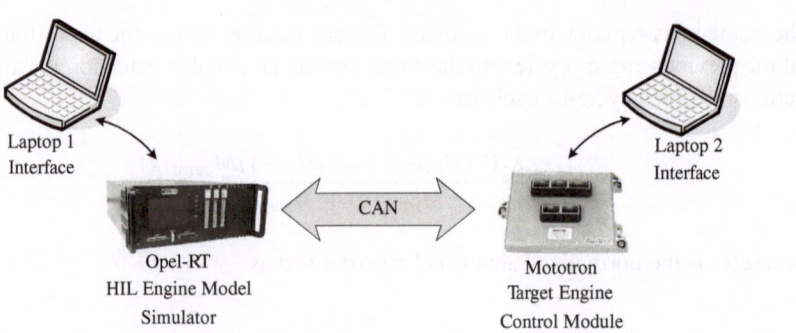

Fig. 4.13 HIL engine model and controller setup

The Opal-RT simulation step size of 1 ms was chosen in order to emulate a real-world continuous time engine. Similarly, the Mototron sample rate of 5 ms for the controller updating is used in many production engine control systems. The CAN communication between Opal-RT and Mototron has a time delay between the time when signals are sent from Mototron and the time when they are received by Opal-RT, and vice versa. This delay was less than 1 ms for our setup since only a few variables were communicated between the HIL simulator and Mototron controller, see the timing scheme in Fig. 4.14. The event-based function call was implemented as follows. At each sample time, the controller checks if the event-based sample condition is met; and if so, the function call will be made to execute the event-based control strategy (see Fig. 4.14). Since the sample period of the event-based LPV controller is a function of engine speed and it can executed with a 5 ms sample period, the LPV controller cannot be updated exactly at each fuel injection event. This leads to some sample time error between ideal event-based sampling and actual function call implementation.

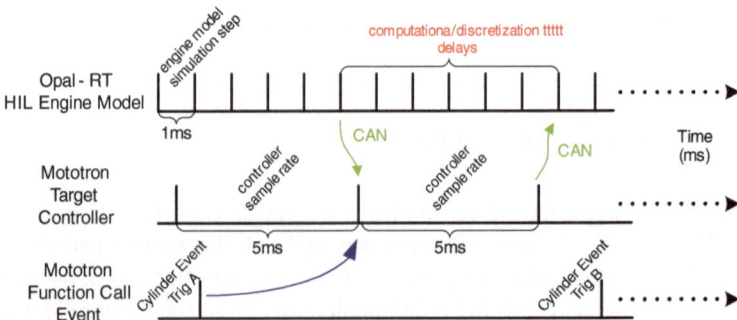

Fig. 4.14 HIL timing scheme

4.7 HIL Simulation Results

In Figs. 4.15, 4.16, 4.17 and 4.18, the responses for the gain-scheduling PI and PID controllers are given by, respectively, the solid gray and black lines. The gray dashed line shows the response of the fixed gain $\mathcal{H}_\infty$ PID controller. In each of the HIL simulations, white Gaussian noise was added to each of the measured signals to represent measurement noise. The standard deviation of the noise added to each signal was set such that the value of the noise would not be larger than the following percentages of the measured signals: air flow $m_A \sim 3\,\%$, equivalence ratio $y_s \sim 2\,\%$, coolant temperature $\sim 5\,\%$, intake pressure $\sim 5\,\%$, and engine speed $N_e \sim 1\,\%$. Even though the cycle-to-cycle combustion variations typically present in internal combustion engines are correlated to engine speed, load, and temperatures, the sensor measurement noise, due to cycle-to-cycle combustion variations and sensor noise, was simplified as a Gaussian white noise due to its simplicity and broad bandwidth. Also, in each of the HIL simulations, the fuel injected is saturated, as a function of the mass air flow, to $\pm 25\,\%$ of the fueling that keeps equivalence ratio at one.

4.7.1 Case 1: Engine Cold Start

We simulate an engine cold start process with coolant temperature at $0\,^\circ\mathrm{C}$ to its normal operational coolant temperature of approximately $100\,^\circ\mathrm{C}$ within about 2 min at an engine speed of $1,500$ rpm. The purpose of this simulation is to emulate the cold start of an internal combustion engine when the engine is operated at high idle speed during the warm-up. Note that during the engine warm-up process the fuel vapor is much less at low temperature than that at high temperature. Therefore, this leads to quite different wall-wetting dynamics. The wall-wetting dynamic coefficients α and β defined in (4.3) were obtained from actual engine test data and they are functions of engine coolant temperature, speed, and load. Since speed and load were fixed in this simulation, both α and β were functions of engine coolant temperature and their values are shown in Fig. 4.16c. The responses of the gain scheduling PI and PID controllers during this simulation, given in Fig. 4.16, are nearly identical. However, at between 100 and 110 s, the fixed gain $\mathcal{H}_\infty$ PID controller becomes saturated causing the measured equivalence ratio to oscillate between 0.8 and 1.2, while both LPV controllers continue to regulate the equivalence ratio to the desired value of 1. Also, in Fig. 4.16b, the mass of the fuel injected when using the fixed gain $\mathcal{H}_\infty$ PID controller has noticeable perturbations due to the noise added to the measured equivalence ratio. However, the gain scheduling PI and PID controllers have no noticeable perturbations which demonstrates that not only do they remain stable over the entire operating range of the engine, but they are also robust to the added measurement noise.

For comparison purposes, a simulation was carried out using the control model described in Sect. 4.2 for the engine cold start problem with the response displayed in

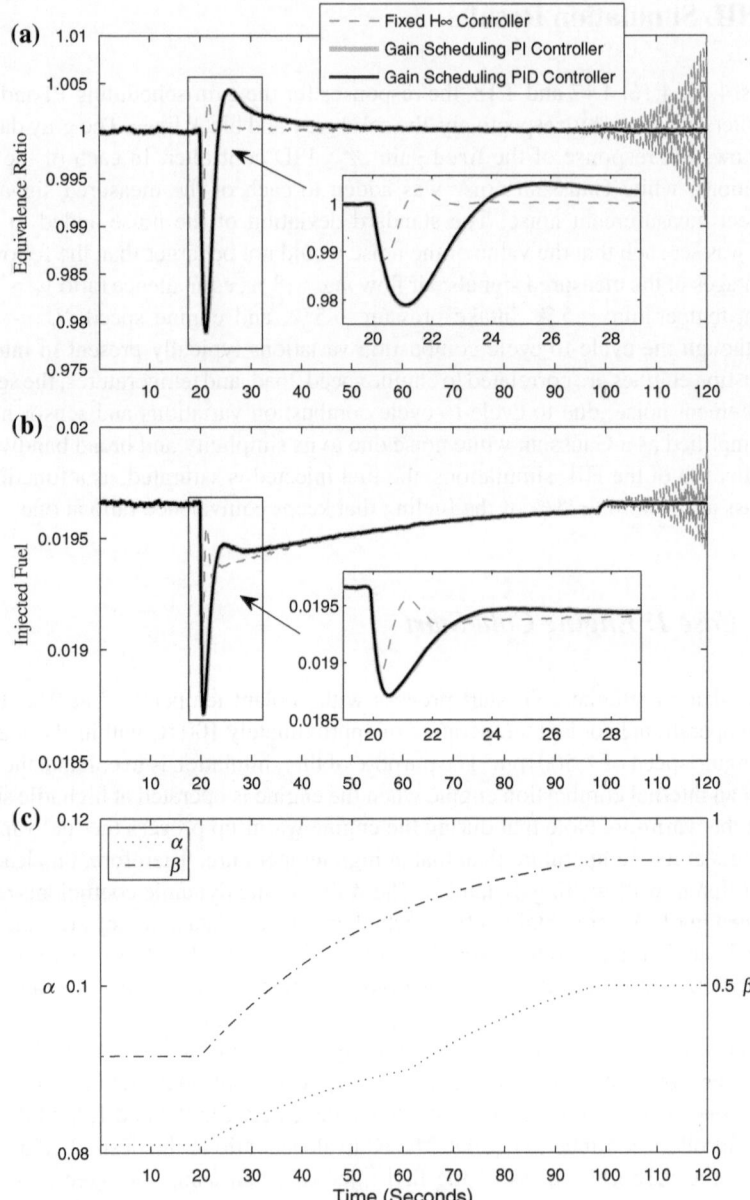

Fig. 4.15 Case 1: engine cold start using simple model

Fig. 4.15. In this simulation, no measurement noise is added to the measured signals. Also, a saturation level is not imposed on the feedback control input.

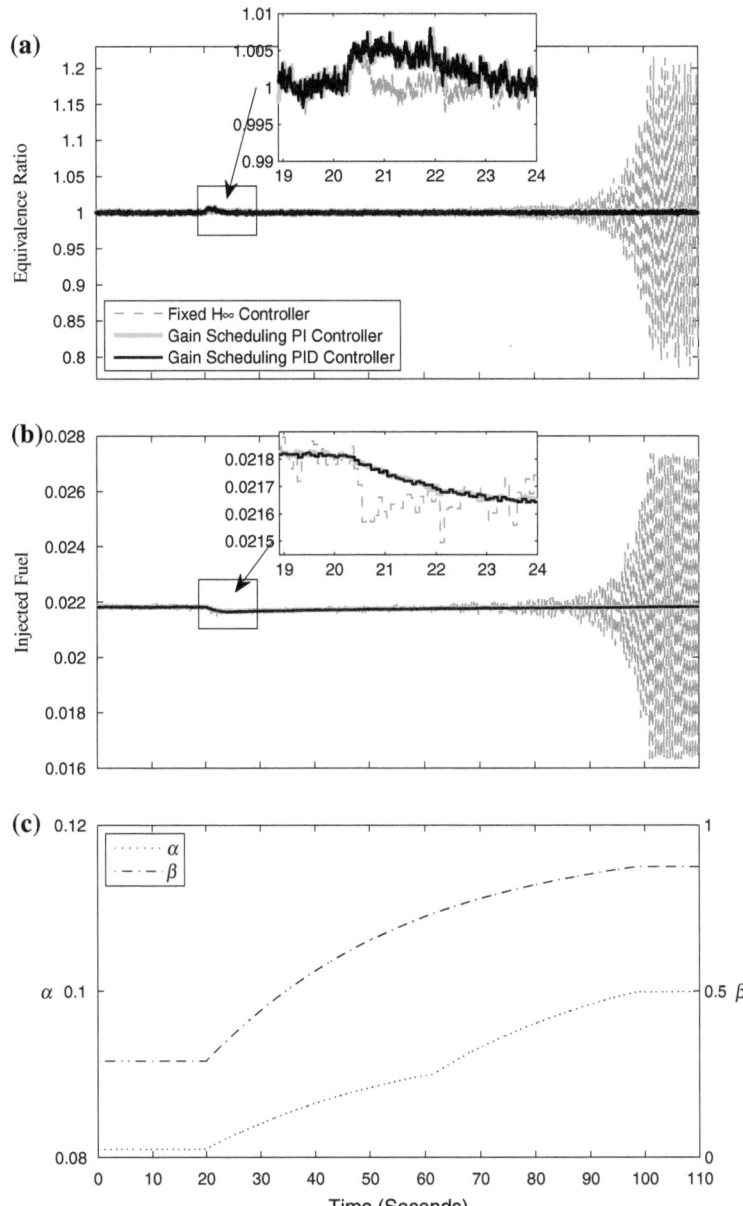

Fig. 4.16 Case 1: engine cold start using HIL

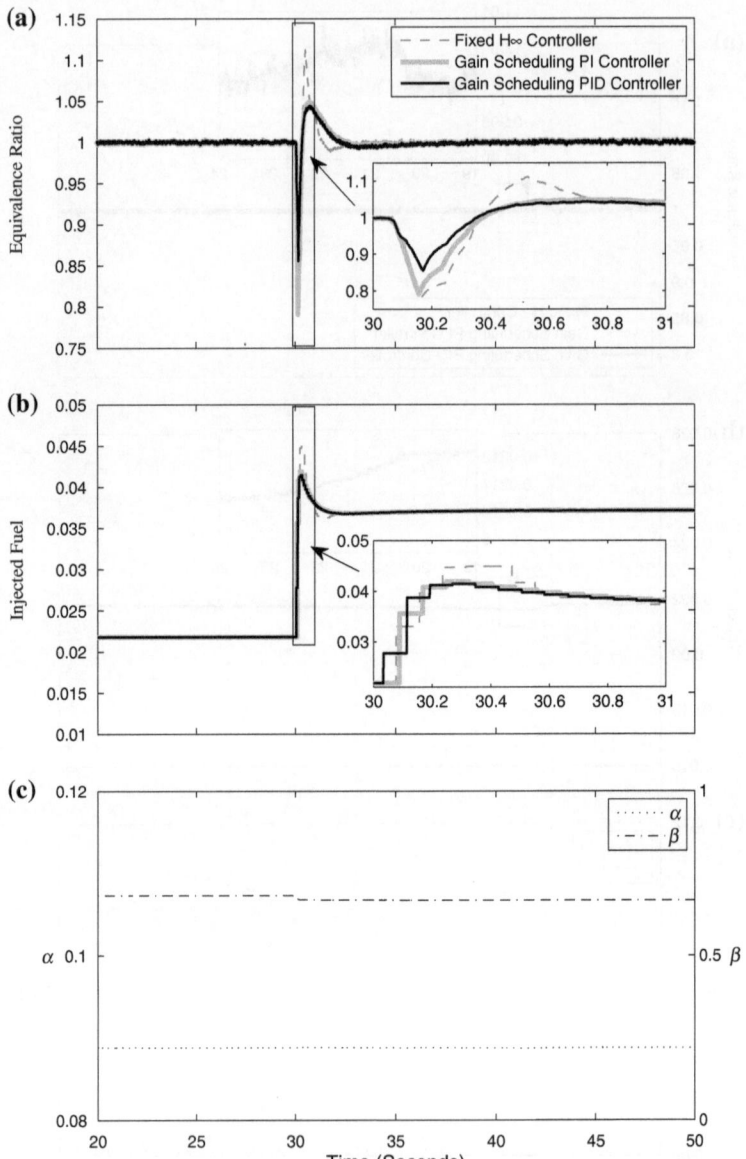

Fig. 4.17 Case 2: engine load change using HIL

4.7.2 Case 2: Load Change

In this case we simulate an engine dynamometer experiment for an engine operated at a coolant temperature of 80 °C with an engine speed of 1,500 rpm. After the engine

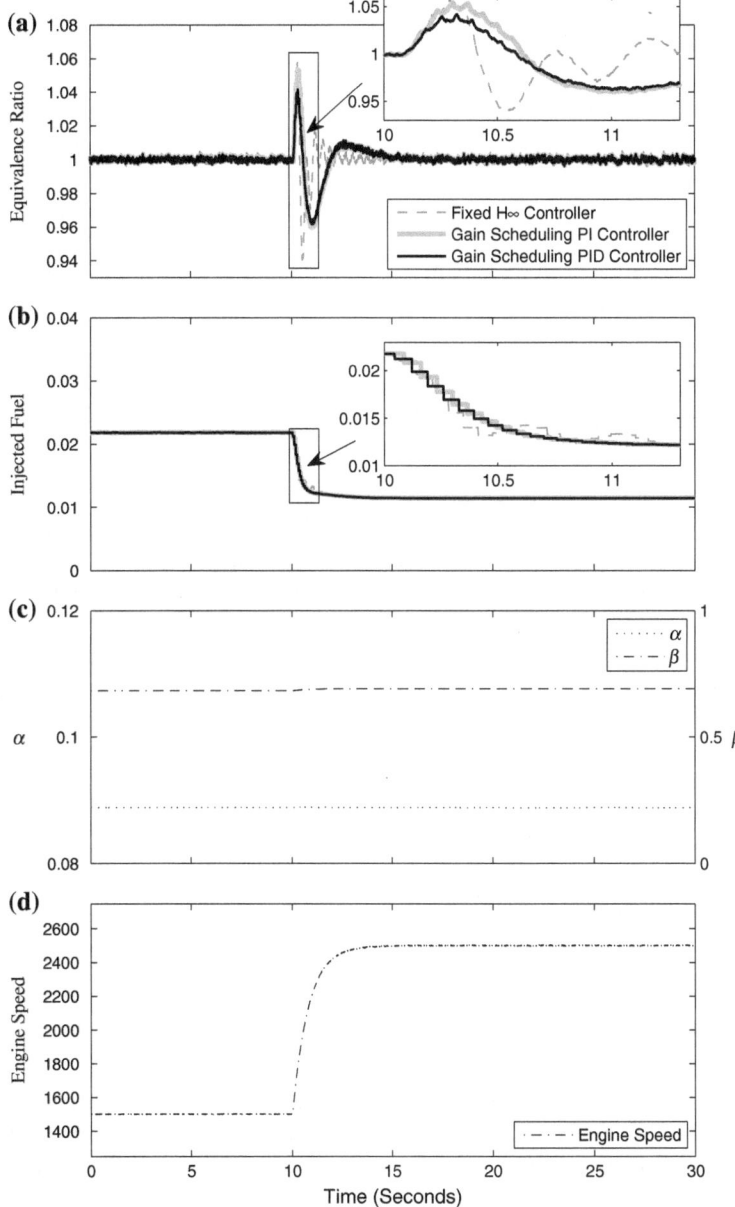

Fig. 4.18 Case 3: engine speed change using HIL

is stably operated at this condition with a 32 % throttle, the load is increased by a step throttle position from 32 to 46 %. Note that in the dynamometer test, the engine speed was maintained by dynamometer through torque regulation. This is similar to the driving condition that a step throttle is applied to maintain the vehicle speed when the vehicle is driven up a hill. Note that the step increment of throttle position produces a slight change in the wall-wetting parameter β as shown in Fig. 4.17c. The responses of each controller is given in Fig. 4.17a. Notice that the throttle step occurring at the 30th second results in a drop in the equivalence ratio due to the step air mass flow. In the detail of Fig. 4.17a, we see that with the gain-scheduling PID controller the equivalence ratio only drops to approximately 0.85, while the gain-scheduling PI and fixed gain $\mathcal{H}_\infty$ controller both drop to nearly 0.8. Also, notice that the equivalence ratio with fixed gain $\mathcal{H}_\infty$ PID controller overshot to over 1.1 with over fueling as seen in the detail of Fig. 4.17b.

4.7.3 Case 3: Engine Speed Change

In this simulation, an engine was operated on a dynamometer with its coolant temperature at 80 °C. To demonstrate the capability for the gain-scheduling controller to handle engine speed variations, a smoothed step command from 1,500 to 2,500 rpm was applied to the engine dynamometer to manipulate the engine speed as shown in Fig. 4.18d. The resulting engine wall-wetting dynamic parameters, shown in Fig. 4.18c, were used in the simulation. Notice in Fig. 4.18a that the gain-scheduling PID controller regulates the equivalence ratio of the engine to the target value of 1 within 5 % error, while the measured equivalence ratio of the engine with the gain-scheduling PI controller and the fixed gain $\mathcal{H}_\infty$ PID controller go above 1.05. Also, the equivalence ratio with the fixed gain $\mathcal{H}_\infty$ PID controller drops to below 0.95, while both gain-scheduling controllers only lower the equivalence ratio to about 0.96. The equivalence ratio with the fixed gain $\mathcal{H}_\infty$ PID controller also has many oscillations and uses more control effort as shown in the detail of Fig. 4.18b, which hurts engine transient fuel economy.

4.7.4 Case 4: Combined Load and Engine Speed Change

In this simulation, an engine was operated on a dynamometer with its coolant temperature at 80 °C. To demonstrate the capability for the gain scheduling controller to handle load changes combined with engine speed variations, the load is increased by a step throttle position from 32 to 46 % and then combined with an engine speed variation generated by a smoothed step command from 1,500 to 2,000 rpm as shown in Fig. 4.19d. The resulting engine wall-wetting dynamic parameters are shown in Fig. 4.19c. Notice in Fig. 4.19a both the gain-scheduling controllers drop the measured equivalence ratio to approximately 0.85, while the fixed gain $\mathcal{H}_\infty$ PID controller drops the measured equivalence ratio below 0.85. Also, the fixed gain $\mathcal{H}_\infty$ PID controller overshoots to nearly 1.15 with over-fueling as seen in the detail of Fig. 4.17b.

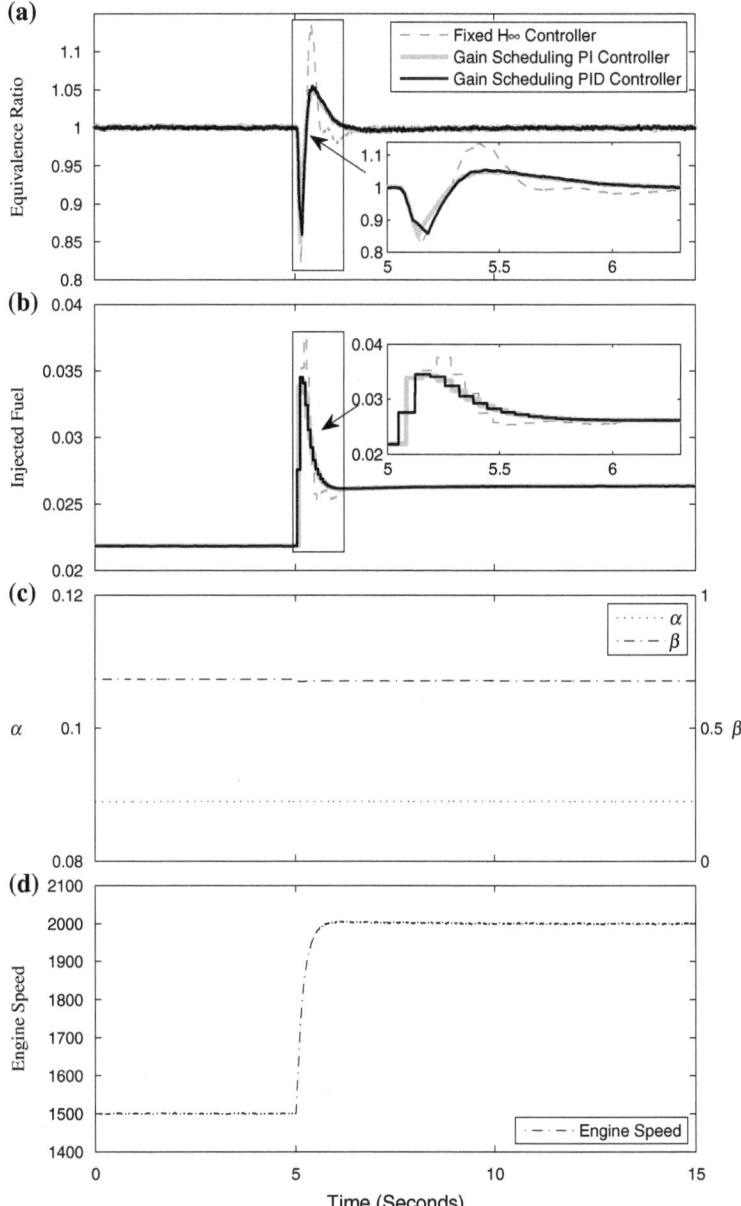

Fig. 4.19 Case 4: combined load and engine speed change using HIL

4.8 Conclusion

In this chapter, a systematic process for developing gain-scheduling PI and PID controllers for discrete-time LPV systems is presented. First, a control-oriented LPV model is developed by using the dynamics of a port-fuel-injection process. Then the LPV model obtained is investigated and found to contain parameter variation that is not affine. Due to limitations in current LPV control schemes for discrete-time systems discussed in Sect. 4.3.3, a first-order Taylor series approximation is performed on the LPV system $H(\Theta)$ in (4.26) to obtain an approximated LPV system $\hat{H}(\Theta)$ in (4.27) with only affine parameter variation. The measurement for control is generated by augmenting the approximated LPV system with a low-pass filter and an integrator. The augmented, approximated LPV system is then converted into a polytopic LPV system so that the synthesis method given by [11] can be utilized. To validate the gain-scheduling controller found with the finally obtained LPV system $\hat{H}(\Theta)$, simulations are performed using the original LPV system $H(\Theta)$. From the simulation results, it is clear that although the approximated LPV system $\hat{H}(\Theta)$ is used to design the gain-scheduling controller it still performs very well when applied to the original LPV system $H(\Theta)$. Furthermore, not only do the HIL simulation results reaffirm the success of the simulation results, they also demonstrate the feasibility of implementing of the proposed LPV scheme on a hardware controller that could be used as an engine control module.

Chapter 5
LPV Control of a Hydraulic Engine Cam Phasing Actuator

In this chapter, an LPV design example [63] that demonstrates how to design a dynamic, output-feedback gain scheduling controller using the LPV methods from Chap. 2 is presented. First, an LPV model is formulated from a family of linear models that were obtained from a series of closed-loop system identification tests for a variable valve timing cam phaser system [49]. Then a control strategy is developed and relevant control structures are appended onto the LPV system to produce the generalized LPV plant. A discussion on weighting function selection for mixed controller synthesis is presented, with an emphasis placed on examining various frequency responses of the system. The generalized LPV plant is then used with the $\mathcal{H}_2/\mathcal{H}_\infty$ LPV controller synthesis presented in Sect. 2.3.3 to synthesize the LPV controller. The LPV controller is then tested with the original variable valve timing cam phaser system for validation.

5.1 Introduction

The intake and exhaust valve timing of an internal combustion (IC) engine greatly influence the fuel economy, emissions, and performance of an IC engine. Conventional valvetrain systems can only optimize the intake and exhaust valve timing for one given operational condition. That is, the optimized valve timing can either improve fuel economy and reduce emissions at low engine speeds or maximize engine power and torque outputs at high engine speeds. However, with the development of continuously variable valve timing (VVT) systems [38], the intake and exhaust valve timing can be modified as a function of engine speed and load to obtain both improved fuel economy and reduced emissions at low engine speeds and increased power and torque at high engine speeds.

To adjust the intake and exhaust timing, the most common cam phasing system is the hydraulic van type cam phaser [21]. The control of hydraulic cam phasing systems has been discussed in [25] and [49]. In [25], a significant nonlinearity in the hydraulic cam phasing system is noted and a nonlinear controller is designed to

A. P. White et al., *Linear Parameter-Varying Control for Engineering Applications*, SpringerBriefs in Control, Automation and Robotics, DOI: 10.1007/978-1-4471-5040-4_5, © The Author(s) 2013

compensate for it. In [49], an $\mathcal{H}_2$ controller is designed using the output covariance constraint (OCC) control design approach [76]. In this chapter, a gain-scheduling controller is developed using an LPV control design.

To obtain the model of the VVT system, closed-loop system identification was used in [48] and [49]. A main reason for selecting closed-loop system identification in [48] and [49] was due to high open-loop gains that makes it difficult to maintain the cam phaser operated at a fixed location for system identification. During the system identification process, it was found that the system gain of the VVT actuator is a function of engine speed, load, oil pressure, and temperature. Therefore, it seems only natural to exploit the knowledge of how the system gain of the VVT actuator varies with the time varying parameters. To do this, the VVT system can be described as a family of linear models to approximate the system dynamics for a given engine speed, load, oil pressure, and temperature. Thus formulating an LPV model for the VVT system.

The purpose of this chapter is to develop a dynamic gain-scheduling controller with guaranteed stability and performance over all time-varying parameters. To do this, the process depicted in Fig. 5.1 was followed. First, a family of LTI models was obtained. Using engine speed and the oil pressure as system parameters, a family of linear models of the VVT system were obtained by performing multiple system identifications while maintaining engine speed and oil pressure at specified levels. With the family of linear models, the LPV model of the VVT system was formulated. To design the dynamic gain-scheduling controller, a standard control structure of observer-based state feedback with integral control was employed. This control structure, along with $\mathcal{H}_2$ and $\mathcal{H}_\infty$ performance weighting functions, were then appended onto the LPV model of the VVT system to obtain the LPV system of

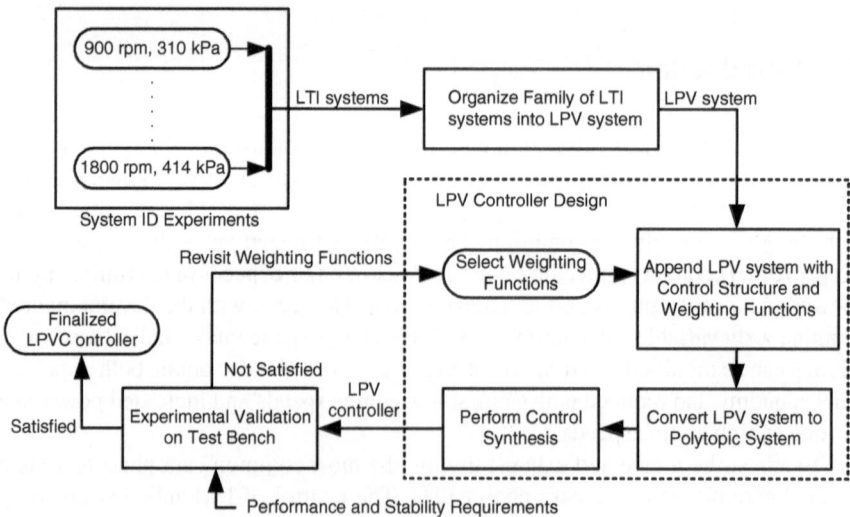

Fig. 5.1 Flowchart of the design and validation process of an LPV controller

the generalized plant. Then the LPV system of the generalized plant was converted to a polytopic system, which is an LPV system with a polytopic dependency on a scheduling parameter that takes values in the unit-simplex, so that the mixed $\mathcal{H}_2/\mathcal{H}_\infty$ discrete-time LPV control synthesis method given by [13] could be applied to obtain the gain-scheduled state feedback and observer gains. Once a potential controller was obtained, its performance was experimentally validated on the test bench used to obtain the family of LTI systems. If the performance and stability requirements of the VVT system are not satisfied when testing the LPV controller, the selected $\mathcal{H}_2$ and $\mathcal{H}_\infty$ performance weighting functions are modified and the control synthesis procedure is performed again. This loop is performed until stability and satisfactory performance are obtained on the test bench.

As stated previously, a multiobjective, mixed $\mathcal{H}_2/\mathcal{H}_\infty$ control design is performed in this chapter. The goal of using both $\mathcal{H}_2$ and $\mathcal{H}_\infty$ performance criteria is to design a controller which can meet multiple performance objectives. In this chapter, a loose $\mathcal{H}_\infty$ performance bound is used to guarantee stability of the closed-loop system under parameter variations. Meanwhile, a tight $\mathcal{H}_2$ performance bound is used to make the LPV controller robust to input disturbances. The selection of $\mathcal{H}_2$ and $\mathcal{H}_\infty$ performance weighting functions is an important design problem. The selection of $\mathcal{H}_\infty$ performance weighting functions can be done as described in [72] and [54]. However, the selection of $\mathcal{H}_2$ performance weighting functions is not covered in such detail. In [76], a systematic way is provided for iteratively tuning the output $\mathcal{H}_2$ weighting functions for robust control of LTI systems. Unfortunately, no such iterative procedure exist yet for LPV systems.

The chapter is organized as follows. The family of linear models obtained from the series of bench identification tests are introduced in Sect. 5.2 and the LPV system is formulated. In Sect. 5.3, the LPV gain-scheduling controller design method is provided. The bench test set-up is discussed in Sect. 5.4.1. In Sect. 5.4.2, the obtained LPV gain-scheduling controller is operated on the test bench and compared to the baseline PI and OCC controllers used in [48]. Concluding remarks are given in the final section.

5.2 LPV System Modeling

To obtain a family of linear models, the closed-loop system identification outlined in [48] was performed at a series of fixed engine speeds N_e and oil pressures p. The open-loop transfer functions of the identified family of linear VVT systems sampled at 5 ms are given by

$$
\begin{aligned}
G(N_e, p = 310\,\text{kPa}\,(45\,\text{psi})) &= \frac{\Psi(N_e, p)\,(0.0859q - 0.0609)}{q^2 - 1.9547q + 0.9553}, \\
G(N_e, p = 414\,\text{kPa}\,(60\,\text{psi})) &= \frac{\Psi(N_e, p)\,(0.0615q - 0.0364)}{q^2 - 1.9547q + 0.9553}
\end{aligned}
\tag{5.1}
$$

Table 5.1 Identified gain $\Psi(N_e, p)$

Pressure, p	Engine speed, N_e (rpm)	Gain $\Psi(N_e, p)$
310 kPa (45 psi)	900	0.70
	1,500	0.72
	1,800	0.68
414 kPa (60 psi)	900	0.95
	1,500	0.98
	1,800	0.93

where $\Psi(N_e, p)$ is the gain at a specific engine speed N_e and oil pressure p as given in Table 5.1 and q is the *forward shift operator* that satisfies $qu(k) = u(k + 1)$.

By inspection of the identified transfer functions in (5.1), the LPV model for the VVT system is given by

$$G(\alpha_k, \beta_k) = \frac{\alpha_k q + \beta_k}{q^2 - 1.9547q + 0.9553} \tag{5.2}$$

where α_k and β_k are used as the time-varying parameters. For notational simplicity, α_k and β_k will be used to denote the parameters at time k, such that $\alpha_k = \alpha(k)$ and $\beta_k = \beta(k)$. The values of α_k and β_k are found for a specific value of engine speed N and oil pressure p by multiplying the appropriate Ψ value found in Table 5.1 with the appropriate transfer function in (5.1). The range of values that α_k and β_k can take are given in Table 5.2.

Using the transfer function in (5.2), a state-space representation of the VVT system is found to be

$$x_G(k + 1) = \underbrace{\begin{bmatrix} 0 & -0.9553 \\ 1 & 1.9547 \end{bmatrix}}_{A_G} x_G(k) + \underbrace{\begin{bmatrix} \beta_k \\ \alpha_k \end{bmatrix}}_{B_G} u_G(k), \tag{5.3}$$

$$y(k) = \underbrace{\begin{bmatrix} 0 & 1 \end{bmatrix}}_{C_G} x_G(k).$$

For convenience, the compact notation $\Theta = [\alpha_k, \beta_k]$ will be used to denote the scheduling variables for the remainder of the chapter.

Table 5.2 Time-varying parameters (scheduling parameters)

$\alpha(N_e(t), p(t)) \in [0.0571, 0.0618]$
$\beta(N_e(t), p(t)) \in [-0.0438, -0.0339]$

5.3 LPV Gain Scheduling Controller Design

5.3.1 Control Strategy

The objective of the control system is to regulate the cam phase y to a reference phase r using feedback control against the disturbance signal d and the time-varying parameters α_k and β_k. In particular, we want to guarantee the stability of the closed-loop system and also minimize the effect of the disturbances for any conceivable engine speed and oil pressure variations. The proposed control architecture is illustrated in Fig. 5.2. This scheme has four components, that is a state observer $\hat{P}(\Theta)$, observer gains $L(\Theta)$, a state feedback controller $K_S(\Theta)$, and an integrator $I(q)$.

The multi-input, single-output LPV plant $P(\Theta)$, depicted inside of the dotted box in Fig. 5.2, is obtained by augmenting the VVT system $G(\Theta)$ with the forward Euler method, discrete-time integrator $I(q) = t_s/(q-1)$, where t_s is the sample period of the discrete-time system in seconds. The integrator $I(q)$ introduces integral action into the system to ensure that the steady-state error between the measured cam phase y and the reference phase r can be eliminated. By allowing the input to the VVT plant $G(\Theta)$ to be equal to

$$u_G(k) = u_P(k) + \frac{t_s}{q-1} u_I(k),$$

as displayed in the dotted box of Fig. 5.2, one possible state-space representation of $P(\Theta)$ is found to be

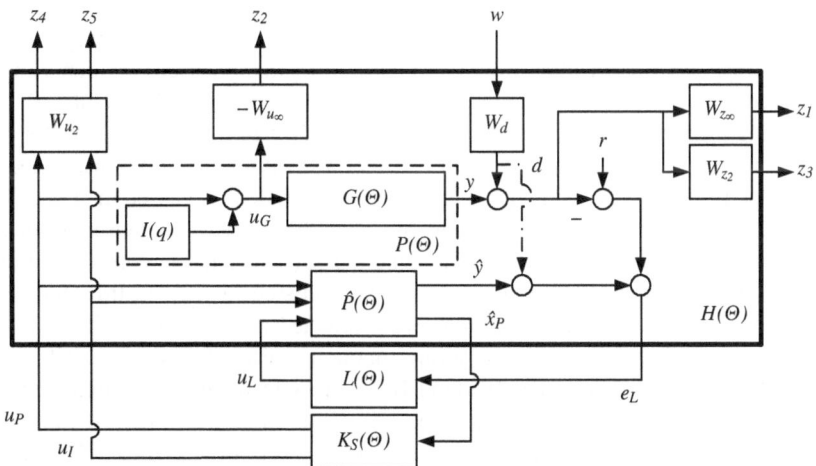

Fig. 5.2 Proposed control architecture for the VVT system

$$x_P(k+1) = \underbrace{\begin{bmatrix} 0 & -0.9553 & \sqrt{t_s}\beta_k \\ 1 & 1.9547 & \sqrt{t_s}\alpha_k \\ 0 & 0 & 1 \end{bmatrix}}_{A_P(\Theta)} x_P(k) + \underbrace{\begin{bmatrix} \beta_k & 0 \\ \alpha_k & 0 \\ 0 & \sqrt{t_s} \end{bmatrix}}_{B_P(\Theta)} \underbrace{\begin{bmatrix} u_P(k) \\ u_I(k) \end{bmatrix}}_{u_S(k)}, \quad (5.4)$$

$$y(k) = \underbrace{\begin{bmatrix} 0 & 1 & 0 \end{bmatrix}}_{C_P} x_P(k).$$

In (5.4), it is clear that the state matrix $A_P(\Theta)$ and the input matrix $B_P(\Theta)$ are both affected by the time-varying parameters α_k and β_k.

The state observer $\hat{P}(\Theta)$ is used to obtain the estimated states $\hat{x}_P$ of the plant. The observer $\hat{P}(\Theta)$ has the standard state-space representation

$$\hat{x}_P(k+1) = A_P(\Theta)\hat{x}_P(k) + B_P(\Theta)u_S(k) + L(\Theta)e_L(k),$$
$$\hat{y}(k) = C_P\hat{x}_P(k),$$

where the error input to the plant observer is given by $e_L(k) = r(k) - (y(k) + d(k)) + (\hat{y}(k) + d(k))$, which simplifies to $e_L(k) = r(k) - y(k) + \hat{y}(k)$. Since we are solving the S/KS mixed-sensitivity $\mathcal{H}_\infty$ optimization using the regulation form, during control synthesis we let the set point r equal zero as shown in [54], thus further simplifying the observer input error to $e_L(k) = -y(k) + \hat{y}(k)$. This satisfies the condition in [13] that the measurement for control is not corrupted by the exogenous input $w(k)$. Notice in Fig. 5.2 that the output disturbance $d(k)$ is connected to the estimated plant output $\hat{y}(k)$ by dash-dot lines. This is to signify that the exogenous input $d(k)$ is only available to the observer during control synthesis. However, during implementation since the output disturbance $d(k)$ cannot be measured it is not available to the observer.

To use mixed $\mathcal{H}_2/\mathcal{H}_\infty$ norms as the performance criteria for shaping the frequency response of the closed-loop system, weighting matrices (which can be considered control design parameters) are introduced in Fig. 5.2. Oftentimes, the weighting matrices are chosen as frequency-dependent functions; however, for this problem static weighting matrices sufficed. The weighting matrix W_d was selected to model the signal d using the signal-based approach discussed in [54]. The $\mathcal{H}_\infty$ performance weighting functions $W_{z\infty}$ and $W_{u\infty}$ were selected to limit the maximum magnitude of the sensitivity function $|S(j\omega)|$ and the controller multiplied by the sensitivity function $|KS(j\omega)|$ as discussed in [72]. In this study, the $\mathcal{H}_\infty$ performance weighting functions were selected primarily for LPV stability. However, the $\mathcal{H}_2$ performance weighting functions were selected for LPV performance. The weighting matrices W_{z2} and W_{u2} were selected using an iterative trial-and-error process. In the iterative process, W_{z2} and W_{u2} started out with values of unity. The control synthesis procedure outlined in Algorithm 1 detailed in Sect. 5.3.4 was then carried out and the sensitivity function was computed and examined. The values used in the weighting function W_{u2} were then increased and the control synthesis was carried out again and the sensitivity function was examined again.

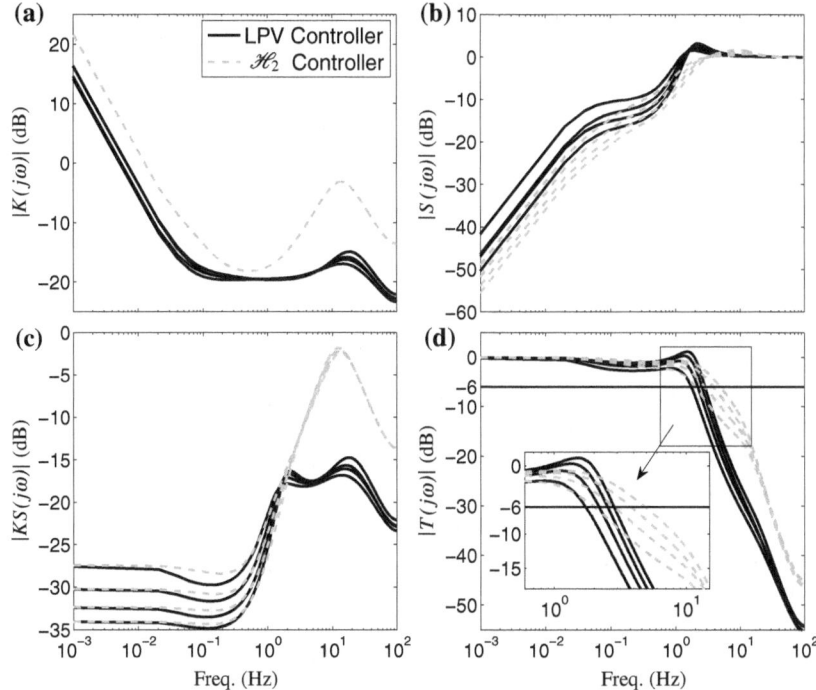

Fig. 5.3 Frequency response comparison of the mixed $\mathcal{H}_2/\mathcal{H}_\infty$ dynamic LPV controller with an OCC ($\mathcal{H}_2$) controller [48] at the corner points of the parameter space polytope

This procedure was executed until desirable characteristics were displayed in the frequency response of the controller, the sensitivity function, and the controller multiplied by the sensitivity function. The resulting weighting matrices are as follows: $W_d = 1$, $W_{u_\infty} = 10$, $W_{u2} = \begin{bmatrix} 15 & 0 \\ 0 & 15 \end{bmatrix}$, $W_{z_\infty} = 1$, and $W_{z2} = 1$. These weighting matrices where tuned to obtain the frequency responses plotted with the bold lines in Fig. 5.3. For comparison, a full-order dynamic output covariance constraint (OCC) controller (dashed lines) [49] was used. This controller is known to work well on the VVT cam phaser test bench at the fixed operational condition of 1,500 rpm and 414 kPa (60 psi) oil pressure, so it was deemed an appropriate starting point.

In Fig. 5.3, the frequency responses of the LPV controller and the OCC controller are displayed at the corner points of the parameter space polytope (i.e. $[\underline{\alpha}, \underline{\beta}]$, $[\underline{\alpha}, \overline{\beta}]$, $[\overline{\alpha}, \underline{\beta}]$, $[\overline{\alpha}, \overline{\beta}]$, where $\underline{\alpha} = \alpha_{\min}$ and $\overline{\alpha} = \alpha_{\max}$). In Fig. ??a, the frequency response of each controller is displayed. At low frequencies, each controller has high gain due to the integral action built into each controller. In Fig. 5.3b, the sensitivity function of each controller is displayed. In a typical feedback system, the sensitivity function is linked to the tracking error performance [72]. At low frequencies, each controller's sensitivity function is small, which minimizes tracking error and maximizes disturbance rejection. Figure 5.3c displays the frequency response of the controller multiplied by the sensitivity function for each controller. This plot shows that over the frequency range of 1–20 Hz the mixed $\mathcal{H}_2/\mathcal{H}_\infty$ dynamic LPV controller has lower control effort than the full-order dynamic OCC controller. Since this is the frequency range over which the output disturbance $d(k)$ is generally active, it means that the mixed $\mathcal{H}_2/\mathcal{H}_\infty$ dynamic LPV controller should be robust to the disturbance $d(k)$. The frequency response of the closed-loop transfer functions with the mixed $\mathcal{H}_2/\mathcal{H}_\infty$ dynamic LPV controller and the OCC controller are displayed in Fig. 5.3d. The benefit of the mixed $\mathcal{H}_2/\mathcal{H}_\infty$ dynamic LPV controller can be seen in the close-up view in Fig. 5.3d. At −6 dB, the closed-loop bandwidth with the OCC controller varies between approximately 2 to 4.8 Hz. However, the closed-loop bandwidth with the LPV controller only varies between approximately 1.8–2.9 Hz, which is a reduction in span of about 60 %.

As displayed in Fig. 5.2, the state feedback gains $K_S(\Theta)$ and the observer gains $L(\Theta)$ are placed outside of the solid, bold box. This designates that the control synthesis in Algorithm 1 is performed on only the items inside of the box. By isolating the static gains $K_S(\Theta)$ and $L(\Theta)$, the design of the observer-based dynamic controller is transformed into the design of a single static controller $K(\Theta)$ by using the following structure:

$$\underbrace{\begin{bmatrix} u_S(k) \\ u_L(k) \end{bmatrix}}_{u(k)} = \underbrace{\begin{bmatrix} K_S(\Theta) & 0 \\ 0 & L(\Theta) \end{bmatrix}}_{K(\Theta)} \underbrace{\begin{bmatrix} \hat{x}_P(k) \\ e_L(k) \end{bmatrix}}_{e(k)} \tag{5.5}$$

where $\hat{x}_p \in \mathbb{R}^s$, $e_L \in \mathbb{R}$, $u_S \in \mathbb{R}^2$, and $u_L \in \mathbb{R}^s$.

5.3.2 Generalized Plant

As shown in Fig. 5.2, the state feedback controller $K_S(\Theta)$ and observer gains $L(\Theta)$ are designed for the generalized LPV plant $H(\Theta)$. The generalized LPV plant $H(\Theta)$ is composed by the multi-input, single-output LPV plant $P(\Theta)$, and its corresponding state observer $\hat{P}(\Theta)$, along with the static weighting matrices W_d, W_{u_∞}, W_{u_2}, W_{z_∞}, and W_{z_2}. The state-space realization of the generalized plant $H(\Theta)$ is found by combining the state-space realizations of $P(\Theta)$ and $\hat{P}(\Theta)$ and performing the connections in Fig. 5.2 to obtain

$$\underbrace{\begin{bmatrix} x_P(k+1) \\ \hat{x}_P(k+1) \end{bmatrix}}_{x(k+1)} = \underbrace{\begin{bmatrix} A_P(\Theta) & 0 \\ 0 & A_P(\Theta) \end{bmatrix}}_{\hat{A}(\Theta)} \underbrace{\begin{bmatrix} x_P(k) \\ \hat{x}_P(k) \end{bmatrix}}_{x(k)} + \underbrace{\begin{bmatrix} B_P(\Theta) & 0 \\ B_P(\Theta) & I \end{bmatrix}}_{\hat{B}(\Theta)} \underbrace{\begin{bmatrix} u_S(k) \\ u_L(k) \end{bmatrix}}_{u(k)}$$

$$z(k) = C_z x(k) + D_w w(k) + D_u u(k) \tag{5.6}$$
$$e(k) = C_e x(k)$$

where $x(k) \in \mathbb{R}^n$ is the state at time k, $w(k) \in \mathbb{R}^r$ is the unweighted exogenous input, $u(k) \in \mathbb{R}^m$ is the control input, $z(k) \in \mathbb{R}^p$ is the performance output, and $e(k) \in \mathbb{R}^q$ is the measurement for control. The state matrix $A_P(\Theta)$ and the input matrix $B_P(\Theta)$ are both given in (5.4) and the other state-space matrices are given by

$$
C_z = \begin{bmatrix} 0 & 1 & 0 & 0 & 0 & 0 \\ 0 & 0 & -10\sqrt{t_s} & 0 & 0 & 0 \\ 0 & 1 & 0 & 0 & 0 & 0 \\ 0 & 0 & 0 & 0 & 0 & 0 \\ 0 & 0 & 0 & 0 & 0 & 0 \end{bmatrix}, \quad D_w = \begin{bmatrix} 1 \\ 0 \\ 1 \\ 0 \\ 0 \end{bmatrix},
$$

$$
D_u = \begin{bmatrix} 0 & 0 & 0 & 0 & 0 \\ -10 & 0 & 0 & 0 & 0 \\ 0 & 0 & 0 & 0 & 0 \\ 15 & 0 & 0 & 0 & 0 \\ 0 & 15 & 0 & 0 & 0 \end{bmatrix}, \quad C_e = \begin{bmatrix} 0 & 0 & 0 & 1 & 0 & 0 \\ 0 & 0 & 0 & 0 & 1 & 0 \\ 0 & 0 & 0 & 0 & 0 & 1 \\ 0 & -1 & 0 & 0 & 1 & 0 \end{bmatrix}.
$$

5.3.3 A Gain-Scheduling Control Synthesis Problem

Now that the state-space representation of the generalized plant $H(\Theta)$ has been obtained, the mixed $\mathcal{H}_2/\mathcal{H}_\infty$ gain-scheduling controller $K(\Theta)$ must be synthesized. The $\mathcal{H}_\infty$-norm from $w(k)$ to $\mathcal{Z}_\infty = [z_1, z_2]^\mathsf{T}$ of the LPV system $H(\Theta)$ in (5.6) with the gain-scheduling controller is defined as

$$
\|H(\Theta)\|_\infty = \sup_{\Theta \in \Theta, w \neq 0} \frac{\|\mathcal{Z}_\infty\|_{\ell_2}}{\|w\|_{\ell_2}}. \tag{5.7}
$$

The $\mathcal{H}_2$-norm from $w(k)$ to $\mathcal{Z}_2 = [z_3(k), z_4(k), z_5(k)]^\mathsf{T}$ of the LPV system $H(\Theta)$ with the gain-scheduling controller is defined as

$$
\|H(\Theta)\|_2^2 = \lim_{T \to \infty} \sup \mathcal{E} \left\{ \frac{1}{T} \sum_{k=0}^{T} \mathcal{Z}_2 \mathcal{Z}_2^\mathsf{T} \right\}, \tag{5.8}
$$

where $\mathcal{E}$ denotes the expectation operator and the positive integer T denotes the time horizon. Now, we formally state the gain-scheduling control design problem.

Problem: The goal is to design a static gain-scheduling control $u(k) = K(\Theta)e(k)$ that stabilizes the closed-loop system and minimizes the worst-case $\mathcal{H}_\infty$ and $\mathcal{H}_2$ norms of the closed-loop LPV system in (5.7) and (5.8) for any trajectories of $\Theta(k) \in \Theta$.

The gain-scheduling method provided by [13] was derived for discrete-time polytopic time-varying systems. Therefore, in the next section, the state-space representation of $H(\Theta)$ in (5.6) will be transformed into a polytopic time-varying system so that the controller $K(\Theta)$ can be synthesized.

5.3.4 Polytopic Linear Time-Varying System

The state-space representation of the generalized plant $H(\Theta)$ in (5.6) can be converted into a discrete-time polytopic time-varying system by solving the state matrix $\hat{A}(\Theta)$ and the input matrix $\hat{B}(\Theta)$ at the vertices of the parameter space polytope, e.g., the state matrix at vertice $\mathcal{V}_2$ is given by $A_2 = \hat{A}(\Theta = [\underline{\alpha}, \overline{\beta}])$. Any Θ inside of the convex parameter set is represented by a convex combination of the vertex systems as weighted by the vector $\lambda(k)$ of barycentric coordinates. Barycentric coordinates are used to specify the location of a point as the center of mass, or barycenter of masses placed at the vertices of a simplex. A formula for computing the barycentric coordinates for any convex polytope is provided by [60]. The discrete-time polytopic time-varying system is given by

$$x(k+1) = A(\lambda(k))x(k) + B_u(\lambda(k))u(k) \qquad (5.9)$$
$$z(k) = C_z x(k) + D_w w(k) + D_u u(k)$$
$$e(k) = C_e x(k)$$

where the state matrix $A(\lambda(k)) \in \mathbb{R}^{n \times n}$ and the input matrix $B(\lambda(k)) \in \mathbb{R}^{n \times m}$ belong to the polytope

$$\mathcal{D} = \left\{ (A, B_u)(\lambda(k)) : (A, B_u)(\lambda(k)) = \sum_{i=1}^{4} \lambda_i(k)(A, B_u)_i, \lambda(k) \in \Lambda \right\}, \quad (5.10)$$

and the other state-space matrices are the same as in (5.6). The state matrix $A(\lambda(k))$ and the input matrix $B_u(\lambda(k))$ are the weighted summation of the vertex matrices as weighted by the vector $\lambda(k)$ of barycentric coordinates, i.e.,

$$A(\lambda(k)) = \sum_{i=1}^{4} \lambda_i(k)A_i \quad \text{and} \quad B_u(\lambda(k)) = \sum_{i=1}^{4} \lambda_i(k)B_i$$

where A_i and B_i are the vertices of the polytope and $\lambda(k) \in \mathbb{R}^4$ is the barycentric coordinate vector which exists in the unit simplex

$$\Lambda = \left\{ \zeta \in \mathbb{R}^4 : \sum_{i=1}^{4} \zeta_i = 1, \zeta_i \geq 0, \quad i = 1, \ldots, 4 \right\}. \qquad (5.11)$$

For all $k \in \mathbb{Z}_{\geq 0}$, the rate of variation of the barycentric coordinates $\Delta\lambda_i(k) = \lambda_i(k+1) - \lambda_i(k)$, is limited such that $-b \leq \Delta\lambda_i(k) \leq b$, with $b \in [0, 1]$, which should be selected with the application in mind. If a worst-case set of parameter variation is known, then this bound can be calculated.

A finite set of LMIs in [13] can be used to design the $\mathcal{H}_2/\mathcal{H}_\infty$ gain-scheduling controller $K(\Theta)$ in (5.5). Due to Theorems 8 and 9 of [13] (see Lemma 2.9), if there exists for $i = 1, \ldots, 4$, matrices $G_{i,K_s} \in \mathbb{R}^{s \times s}$, $G_{i,L} \in \mathbb{R}^{(q-s) \times (q-s)}$, $Z_{i,K_s} \in \mathbb{R}^{(m-s) \times s}$, and $Z_{i,L} \in \mathbb{R}^{s \times 1}$ assembled as

$$G_{i,1} = \begin{bmatrix} G_{i,K_s} & 0 \\ 0 & G_{i,L} \end{bmatrix} \quad \text{and} \quad Z_{i,1} = \begin{bmatrix} Z_{i,K_s} & 0 \\ 0 & Z_{i,L} \end{bmatrix}, \quad (5.12)$$

along with the other matrix variables defined in Lemma 2.9, then the $\mathcal{H}_2/\mathcal{H}_\infty$ controller $K(\lambda(k))$ is given by

$$K(\lambda(k)) = \hat{Z}(\lambda(k))\hat{G}(\lambda(k))^{-1}, \quad (5.13)$$

with

$$\hat{Z}(\lambda(k)) = \sum_{i=1}^{4} \lambda_i(k) Z_{i,1} \quad \text{and} \quad \hat{G}(\lambda(k)) = \sum_{i=1}^{4} \lambda_i(k) G_{i,1}.$$

This control is proved to stabilize affine parameter-dependent systems such as (5.9) with a guaranteed $\mathcal{H}_2$ and $\mathcal{H}_\infty$ performance for all $\lambda \in \Lambda$ and $|\Delta\lambda| \leq b$. In this work, to ensure that all possible parameter variations would be covered, we selected $b = 0.4$. The LMI conditions of Theorems 8 and 9 of [13] are solved by programming them into MATLAB using the LMI parser YALMIP [35] and solved using SeDuMi [56]. During the solution process, the goal is to calculate the gain-scheduled feedback controller $K(\lambda(k))$ that minimizes the bound ν on the $\mathcal{H}_2$ performance from $w(k)$ to $[z_3, z_4, z_5]^T$ under a prescribed bound η on the $\mathcal{H}_\infty$ norm from $w(k)$ to $[z_1, z_2]^T$. The procedure for performing the mixed $\mathcal{H}_2/\mathcal{H}_\infty$ control synthesis is outlined in Algorithm 1.

Note that the minimum feasible $\mathcal{H}_\infty$ bound η_L can be solved for by using an iterative algorithm [54], such as the bisection algorithm.

The resulting LPV controller solved at an engine speed and oil pressure of $N_e = 1{,}500$ rpm and $p = 414$ kPa (60 psi) (for comparison with the $\mathcal{H}_2$ output covariance controller) is found to be

$$K_{\text{LPV}}(q) = \frac{0.109255247q^3 - 0.302866405q^2 + 0.278279285q - 0.0846677044}{q^4 - 3.132121334q^3 + 3.625898107q^2 - 1.853079890q + 0.359303117}. \quad (5.14)$$

As stated previously, the robust $\mathcal{H}_2$ controller designed in [49] using the OCC control design algorithm presented in [76] is used for comparison with the LPV controller. The robust $\mathcal{H}_2$ OCC controller designed in [49] is given by

$$K_{\text{OCC}}(q) = \frac{0.3158302q^3 - 0.9301618q^2 + 0.9129406q - 0.2986088}{q^4 - 3.4051293q^3 + 4.3533113q^2 - 2.4909563q + 0.5427743}. \quad (5.15)$$

Algorithm 1 Mixed $\mathcal{H}_2/\mathcal{H}_\infty$ Gain-Scheduling Synthesis

---

Require: Polytopic LPV system in (5.9), rate of variation bound $b \in [0, 1]$, $\mathcal{H}_2$ and $\mathcal{H}_\infty$ input and output channels of (5.9), and a range of prescribed $\mathcal{H}_\infty$ bounds $\eta \in [\eta_L, \eta_U]$, where it is assumed that η_L is the minimum feasible $\mathcal{H}_\infty$ bound.

Ensure: The gain-scheduling controller matrices $G_{i,1}$ and $Z_{i,1}$ needed to compute $K(\lambda(k))$ in (5.13).

1: Determine selection matrices L_j and M_j for each performance specification j as in Sect. 5.3 of [13].

2: Compute H_j using selection matrices L_j and M_j for each performance specification j, for $j = 1, 2$.

3: Compute the vectors f^j and h^j using rate of variation bound b as shown in Appendix 11.3 of [13].

4: Using equation (29) of [13], convert the polytopic LPV system in (5.9) to the form used in the LMIs of Theorems 8 and 9 of [13].

5: **for** $\eta = \eta_L : \eta_U$ **do**

6:   Initialize the matrix variables introduced in Theorems 8 and 9 of [13] as free matrix variables into MATLAB using the YALMIP interface [35].

7:   Using G_{i,K_s}, $G_{i,L}$, Z_{i,K_s}, and $Z_{i,L}$, generate $G_{i,1}$ and $Z_{i,1}$ as shown in (5.12).

8:   Using the YALMIP interface [35], program the $\mathcal{H}_\infty$ LMIs in Theorem 8 of [13] using prescribed bound η and the $\mathcal{H}_2$ LMIs in Theorem 9 of [13] into MATLAB.

9:   Using an LMI solver, like SeDuMi [56], solve the system of LMIs with the objective of minimizing $\sum_{i=1}^{4} \text{trace}\{W_i\}$, where W_i is a positive-definite $\mathcal{H}_2$ free matrix variable introduced in Theorem 9 of [13], thus minimizing the $\mathcal{H}_2$ norm.

10: **end for**

11: Select the solution that minimizes the $\mathcal{H}_2$ norm the most, yet still has an acceptable bound η on the $\mathcal{H}_\infty$ norm.

5.4 VVT System Test Bench

5.4.1 Bench Test Set-Up

The closed-loop system identification outlined in Ref. [49] and the control design testing were conducted on the VVT test bench displayed in Fig. 5.4. A Ford 5.4L V8 engine head was modified and mounted on the test bench. The cylinder head has a single cam shaft with a VVT actuator for one exhaust and two intake valves. These valves introduce a cyclic torque disturbance to the cam shaft. The cam shaft is driven by an electrical motor (simulating the crankshaft) through a timing belt, see Fig. 5.5.

An encoder is installed on the motor shaft, which generates the crank angle signal with one degree resolution, along with a so-called gate signal (one pulse per revolution). A plate with five magnets adhered is mounted at the other side of the extended cam shaft. As displayed in Fig. 5.5, one magnet is placed on the edge of the plate and is used to synchronize the top dead center position of the combustion phase. The other four magnets on the face of the plate are used to determine the cam phase four times per engine cycle. The two squares in Fig. 5.5 represent hall-effect cam position sensors. As the cam shaft rotates, the magnets on the plate face pass the hall-effect cam position sensor used to determine cam phase and the magnet on

Fig. 5.4 VVT phase actuator test bench

Fig. 5.5 VVT phase actuator
test bench diagram

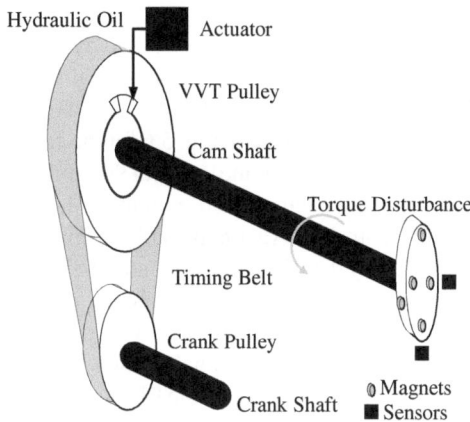

the edge of the plate passes the hall-effect cam position sensor used to determine top dead center position. Within an engine cycle, the cam position sensor generates four cam position pulses, which are sampled by an Opal-RT real-time controller. By comparing these pulse locations with respect to the encoder gate signal, the Opal-RT controller calculates the cam phase with one crank degree resolution four times per an engine cycle.

The cam phase actuator system consists of a solenoid driver circuit, a solenoid actuator, and a hydraulic cam actuator. The solenoid actuator is controlled by a pulse-width modulation (PWM) signal, whose duty cycle is linearly proportional to the DC voltage command. An electrical oil pump was used to supply pressurized engine oil to be used for lubrication and as hydraulic actuating fluid for the cam phase actuator. The cam actuator command voltage signal is generated by the Opal-RT prototype controller and sent to the solenoid driver. The PWM duty cycle is linearly proportional to input voltage with a maximum duty cycle 99 % corresponding to 5 V and a minimal duty cycle of 1 % corresponding to 0 V. The solenoid actuator controls the hydraulic fluid (engine oil) flow and changes the cam phase. The cam position sensor signal is sampled by the Open-RT prototype controller and the corresponding cam phase is calculated within the Opal-RT real-time controller.

A PI controller was tuned for the VVT system on the test bench for comparison purpose with the LPV and OCC controllers. The PI gains tuning process was completed at different engine speeds and oil pressures. The following tuned PI controller achieves good balance between response time and over-shoot oscillations at different conditions:

$$K_{PI}(q) = \frac{0.2q - 0.1995}{q - 1}. \tag{5.16}$$

5.4.2 Bench Test Results

The mixed $\mathscr{H}_2/\mathscr{H}_\infty$ observer-based dynamic LPV controller was tested on the VVT cam phaser bench at engine speeds of 900, 1,200, 1,500, and 1,800 rpm for both engine oil pressures of 310 (45) and 414 kPa (60 psi). The step response of each controller is displayed in Fig. 5.6 for the cam advance ($-20°$ to $0°$) and the cam retard ($0°$ to $-20°$) at an engine speed of 900 rpm with an oil pressure of 310 kPa (45 psi). In Fig. 5.6b, the control effort of both the LPV and $\mathscr{H}_2$ controllers is visibly lower than the PI controller. Also noticeable in Fig. 5.6b is that the control effort corrections produced by the LPV controller are smaller than those produced by the $\mathscr{H}_2$ controller. This was anticipated from frequency response plot of each controller in Fig. 5.3a. Since the LPV controller has lower gain than the $\mathscr{H}_2$ controller, it is less sensitive to the change in error signal (which has the resolution of one crank degree in the experiment), which makes the LPV controller more robust to disturbances in the cam phase when compared to the $\mathscr{H}_2$ controller. This is even more noticeable during cam retard operation in Fig. 5.6c, d. The performance of the LPV controller in comparison with the $\mathscr{H}_2$ and PI controllers can also be shown by computing the control variance once the cam phase has reached steady state. During cam advance with an engine speed of 900 rpm and oil pressure of 310 kPa (45 psi), the control variances of the LPV, $\mathscr{H}_2$, and PI controllers were found to be 0.0048 V^2, 0.0265 V^2 and 0.0079 V^2, respectively. During cam retard at the same engine conditions, the control variances of the LPV, $\mathscr{H}_2$, and PI controllers were found to be 0.0063 V^2, 0.0281 V^2 and 0.0068 V^2, respectively. Similar values for the control variance for each controller were found at all other engine conditions tested as well. The control variances of the LPV controller under all engine conditions tested were found to be approximately anywhere from 6 to 33 % of the control variance of the $\mathscr{H}_2$ controller.

In Fig. 5.7, the mean of the measured overshoot from ten test runs at each engine condition is plotted for each controller. It is easy to see from Fig. 5.7a, b, that in all cases both the $\mathscr{H}_2$ controller and LPV controller obtain lower overshoot than the PI controller, with the $\mathscr{H}_2$ controller displaying the lowest overshoot in most cases. However, during the cam retard situation displayed in Fig. 5.7b, the overshoot of the LPV controller is much closer to that of the $\mathscr{H}_2$ controller and is even smaller than

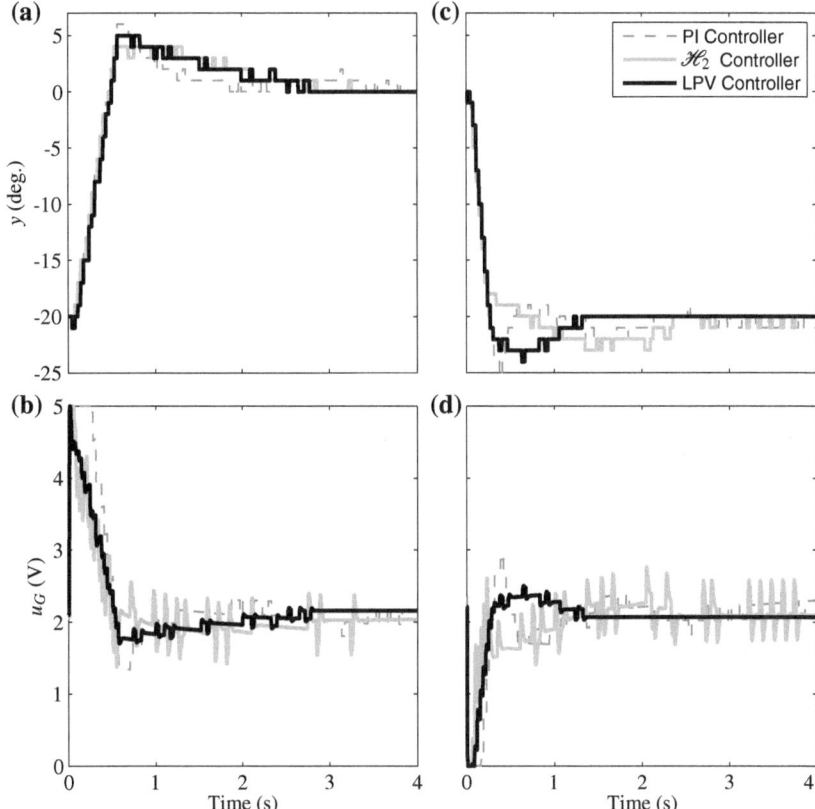

Fig. 5.6 Cam advance response at 900 rpm with 310 kPa oil pressure

the $\mathcal{H}_2$ controller at an engine speed of 1,800 rpm. The difference in performance between cam advance and cam retard is attributed to the fact that the dynamics are slightly different. During cam advance, the actuating torque generated by the oil pressure overcomes the cam load torque causing the cam phase to advance. However, during cam retard, the oil trapped inside the actuator bleeds back to the oil reserve when the cam phase is pushed back by the cam load shaft. This difference in dynamics between the cam advance and cam retard, as shown in Fig. 5.7a, b, generally results in lower overshoot and faster settling and rising times for the cam retard performance compared to the cam advance performance. Also, while the overshoot performance of all of the controllers in Fig. 5.7a, b is above 15 %, none of the controllers include feedforward control. With feedforward control, the overshoot would be significantly reduced.

In Fig. 5.8a, b the mean of the measured 5 % settling time from ten test runs is displayed. It is observed that for nearly all cases, the LPV controller settles quicker than the $\mathcal{H}_2$ controller, with one exception of when the engine is operated with an

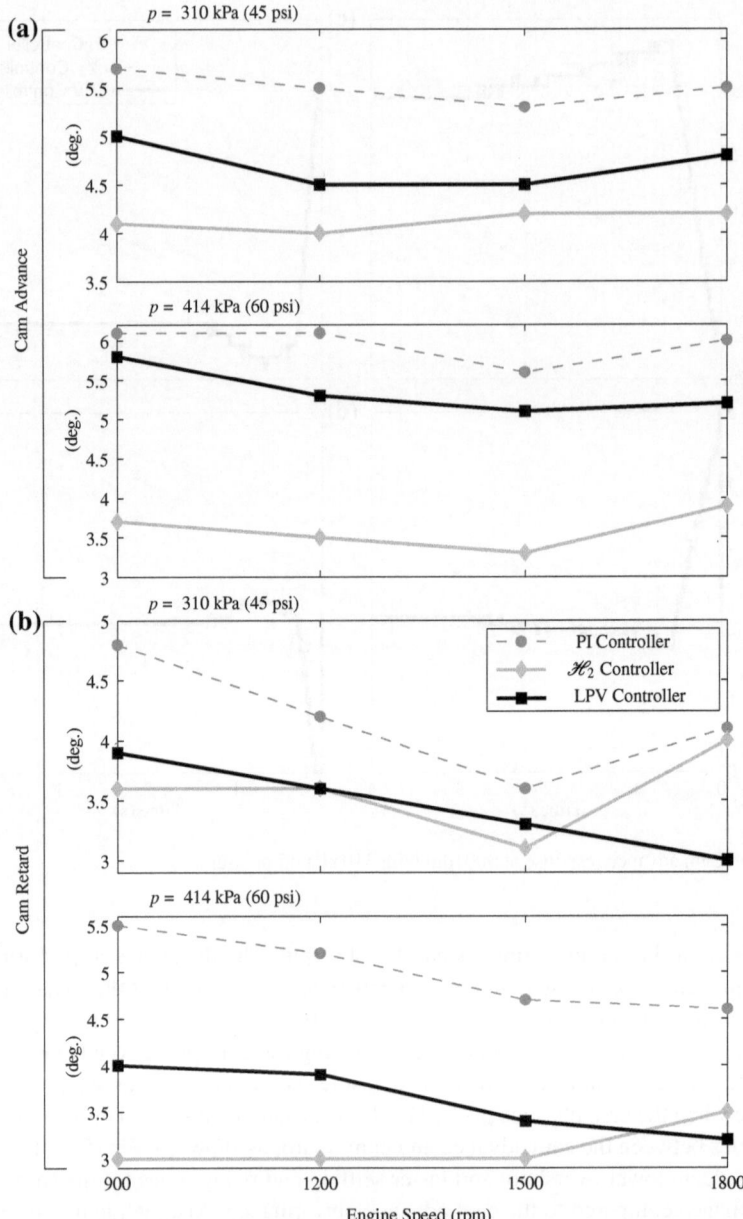

Fig. 5.7 Mean overshoot for each controller operated at an oil pressure of 45 and 60 psi and engine speeds of 900, 1,200, 1,500, and 1,800 rpm. Plot **a** displays the mean overshoot for cam advance and plot **b** displays the mean overshoot for cam retard

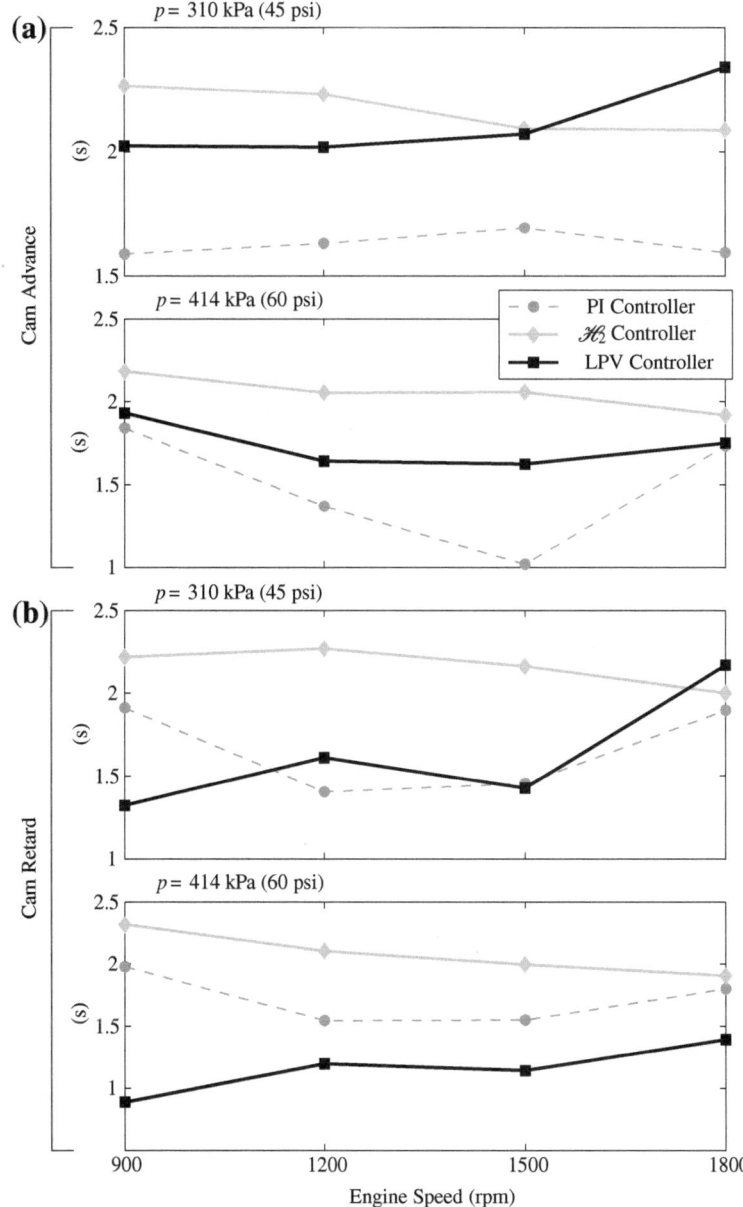

Fig. 5.8 Mean 5 % settling time for each controller operated at an oil pressure of 45 and 60 psi and engine speeds of 900, 1,200, 1,500, and 1,800 rpm. Plot **a** displays the mean settling time for cam advance and plot **b** displays the mean settling time for cam retard

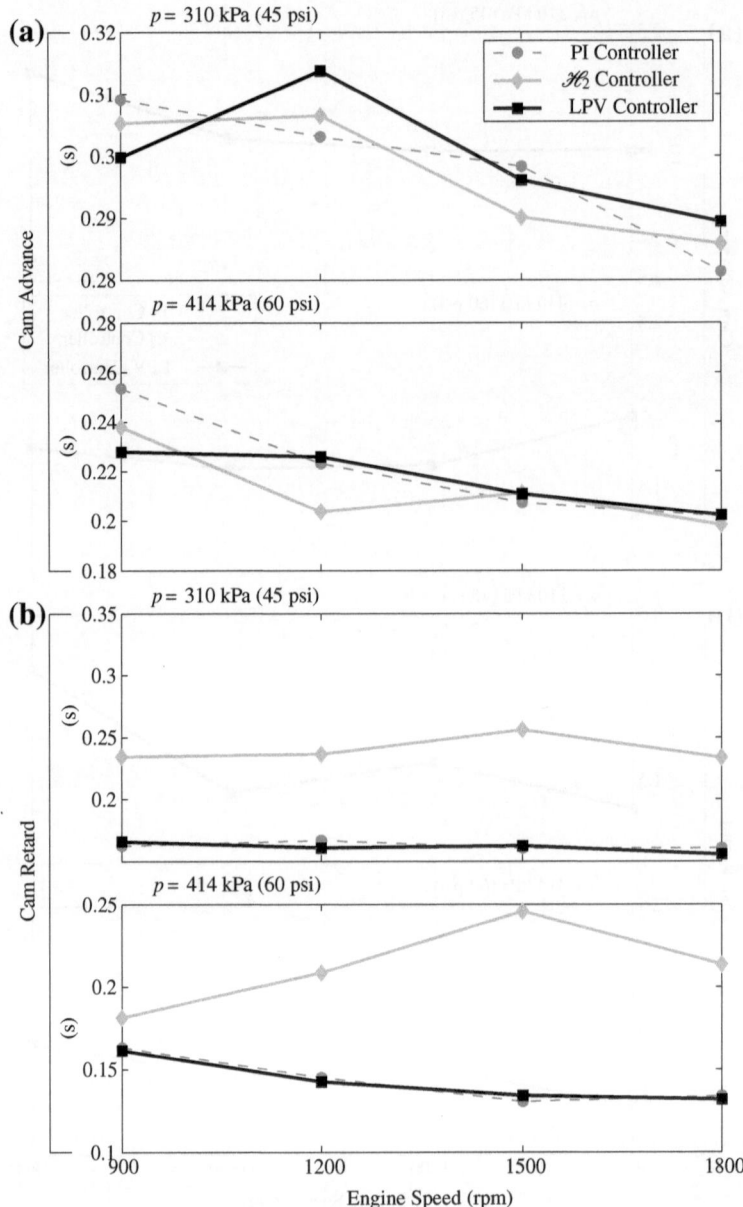

Fig. 5.9 Mean 10–90 % rising time for each controller operated at an oil pressure of 45 and 60 psi and engine speeds of 900, 1,200, 1,500, and 1,800 rpm. Plot **a** displays the mean rising time for cam advance and plot **b** displays the mean rising time for cam retard

oil pressure of 310 kPa (45 psi) and at an engine speed of 1,800 rpm. For the cam advance, the PI controller almost uniformly has the quickest settling time. However, as observed in Fig. 5.8b, during cam retard the settling time of the LPV controller is quicker than the PI controller in most cases, especially when the engine oil pressure is 414 kPa (60 psi).

The mean 10–90 % rising time from the ten test runs is displayed in Fig. 5.9a, b. The rising time performance during cam advance is very similar for each of the controllers as displayed in Fig. 5.9a. However, as shown in Fig. 5.9b, during cam retard it is quite clear that the LPV and PI controllers are faster than the $\mathcal{H}_2$ controller by an unmistakable amount.

5.5 Conclusion

In this chapter, a dynamic gain-scheduling controller was designed by employing an observer-based state feedback design and static multiobjective $\mathcal{H}_2/\mathcal{H}_\infty$ controller synthesis. By examining the frequency response of the LPV controller and comparing it to a previously obtained robust $\mathcal{H}_2$ OCC controller, the LPV controller was found to reduce the operating bandwidth variation of the closed-loop system by approximately 60 %. The frequency response of each system also demonstrated that the LPV controller had lower control effort over the crucial frequency range of 1–20 Hz. This was validated by the bench tests run with each controller, which showed that the LPV controller had much lower control variance than the robust $\mathcal{H}_2$ OCC controller. Also, while the LPV controller is more complex than the PI controller in both concept and implementation, it has lower overshoot than the PI controller at all operating conditions with similar settling and response time characteristics. Additionally, the LPV controller was designed with a systematic approach while the PI controller was obtained through ad hoc testing.

Appendix A
Linear Fractional Transformation

For completeness, we will now give the definition of a linear fractional transformation (LFT). Linear fractional transformations are used to efficiently formulate the interconnection of multi-input multi-output subsystems with multiple sources, such as uncertainties, noises, disturbances, and varying parameters. As given by [72], the possibly complex coefficient matrix M is partitioned as

$$M = \begin{bmatrix} M_{11} & M_{12} \\ M_{21} & M_{22} \end{bmatrix} \in \mathbb{C}^{(p_1+p_2) \times (q_1+q_2)}, \tag{A.1}$$

with $\Delta_\ell \in \mathbb{C}^{q_2 \times p_2}$ and $\Delta_u \in C^{q_1 \times p_1}$. A lower LFT (Fig. A.1a) is given with respect to Δ_ℓ as

$$\mathscr{F}_\ell(M, \Delta_\ell) = M_{11} + M_{12}\Delta_\ell(I - M_{22}\Delta_\ell)^{-1}M_{21}. \tag{A.2}$$

An upper LFT (Fig. A.1b) is given with respect to Δ_u by

$$\mathscr{F}_u(M, \Delta_u) = M_{22} + M_{21}\Delta_u(I - M_{11}\Delta_u)^{-1}M_{12}. \tag{A.3}$$

From the diagrams in Fig. A.1, the reason behind the terminology of lower and upper LFTs should be clear. The set of equations representing the lower LFT diagram in Fig. A.1a are given by

$$\begin{bmatrix} z_1 \\ y_1 \end{bmatrix} = \begin{bmatrix} M_{11} & M_{12} \\ M_{21} & M_{22} \end{bmatrix} \begin{bmatrix} w_1 \\ u_1 \end{bmatrix},$$
$$u_1 = \Delta_\ell y_1, \tag{A.4}$$

and the equations representing Fig. A.1b are given by

$$\begin{bmatrix} y_2 \\ z_2 \end{bmatrix} = \begin{bmatrix} M_{11} & M_{12} \\ M_{21} & M_{22} \end{bmatrix} \begin{bmatrix} u_2 \\ w_2 \end{bmatrix},$$
$$u_2 = \Delta_u y_2. \tag{A.5}$$

A. P. White et al., *Linear Parameter-Varying Control for Engineering Applications*, SpringerBriefs in Control, Automation and Robotics, DOI: 10.1007/978-1-4471-5040-4, © The Author(s) 2013

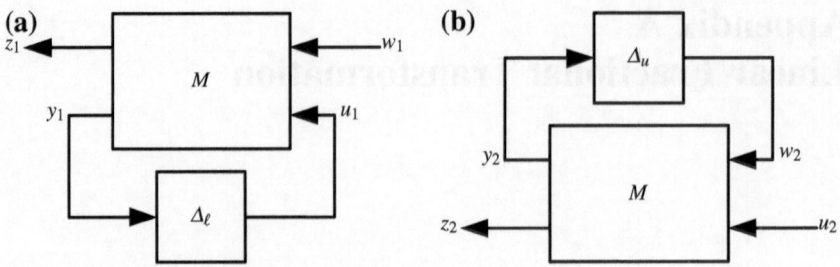

Fig. A.1 **a** Diagram of a *lower* LFT. **b** Diagram of an *upper* LFT

The partitioning of M depends on the interconnections with the isolated parameter Δ_ℓ or Δ_u and can be determined using the MATLAB function "sysic" [6].

Appendix B
Port Fuel Injection System Matrices

The state-space matrices for the LPV system in (4.23) have been found to be

$$
A = \begin{bmatrix}
0.91 & 0 & 0.0369 & 0 & 0 & 0 & 0 & 0 \\
0.2617 & 0 & 0.1544 & 0 & 0 & 1.4352 & 0 & 0 \\
0 & 0 & 0.8475 & 0 & 0 & 0 & 0 & 0 \\
0 & 1.4506 & 0 & 0.2231 & 0.3972 & 0 & 0 & 0 \\
0 & 2.6311 & 0 & 0 & 0.3114 & 0 & 0 & 0 \\
0 & 0 & 0 & 0 & 0 & 0.9986 & 0 & 0 \\
0 & 0 & 0 & 0 & 0 & 0 & 1.9972 & -0.9985 \\
0 & 0 & 0 & 0 & 0 & 0 & 0.9987 & 0
\end{bmatrix} \in \mathbb{R}^{8 \times 8}
$$

(B.1)

$$
B_0 = \begin{bmatrix}
-0.09 & 0.0625 & 0 & 0 & 1 & 0 & 0 \\
0.2617 & 0.2617 & 0 & 0 & 0 & 0 & 0 \\
0 & -0.2585 & 1.6949 & 1.6949 & 0 & 0 & 0 \\
0 & 0 & 0 & 0 & 0 & 0.0664 & -0.0027 \\
0 & 0 & 0 & 0 & 0 & 0.0436 & -0.0073 \\
0 & 0 & 0 & 0 & 0 & 0 & 0 \\
0 & 0 & 0 & 0 & 0 & 0 & 0 \\
0 & 0 & 0 & 0 & 0 & 0 & 0
\end{bmatrix}
$$

$$
\begin{bmatrix}
0 & 0 & 0 & 0 & 0 & 0 & 0 \\
0 & 0 & 0 & 0 & 0 & 0 & 0 \\
0 & 0 & 0 & 0 & 0 & 0 & 0 \\
-0.0214 & -0.0179 & -0.0933 & -0.4891 & -0.0984 & 0.0608 & 0.0975 \\
-0.0186 & -0.0134 & 0 & -0.7266 & 0.1211 & 0.3095 & 0.2231 \\
0 & 0 & 0 & 0 & 0 & 0 & 0 \\
0 & 0 & 0 & 0 & 0 & 0 & 0 \\
0 & 0 & 0 & 0 & 0 & 0 & 0
\end{bmatrix} \in \mathbb{R}^{8 \times 14}
$$

(B.2)

A. P. White et al., *Linear Parameter-Varying Control for Engineering Applications*,
SpringerBriefs in Control, Automation and Robotics,
DOI: 10.1007/978-1-4471-5040-4, © The Author(s) 2013

$$B_1 = \begin{bmatrix} 0 & 0 & 0.0043 \\ 0 & 0 & 0.0179 \\ 0 & 0 & -0.0073 \\ 0 & 0 & 0 \\ 0 & 0 & 0 \\ 0.3756 & 0 & 0 \\ 0 & 0.0266 & 0 \\ 0 & 0 & 0 \end{bmatrix} \in \mathbb{R}^{8 \times 3}, \tag{B.3}$$

$$B_2 = \begin{bmatrix} 0.0369 \\ 0.1544 \\ 0 \\ 0 \\ 0 \\ 0 \\ 0 \\ 0 \end{bmatrix} \in \mathbb{R}^{8 \times 1} \tag{B.4}$$

$$C_0 = \begin{bmatrix} 0 & 0 & 1 & 0 & 0 & 0 & 0 & 0 \\ 0 & 0 & 0 & 0 & 0 & 0 & 0 & 0 \\ 0 & 0 & 0.1525 & 0 & 0 & 0 & 0 & 0 \\ 0 & 0 & -1 & 0 & 0 & 0 & 0 & 0 \\ -1 & 0 & 0.41 & 0 & 0 & 0 & 0 & 0 \\ 0 & 63.6832 & 0 & 0 & 0 & 0 & 0 & 0 \\ 0 & 0 & 0 & 0 & 0 & 0 & 0 & 0 \\ 0 & 0 & 0 & 0 & 0 & 0 & 0 & 0 \\ 0 & 0 & 0 & 0 & 0 & 0 & 0 & 0 \\ 0 & 25.2968 & 0 & 0 & 0 & 0 & 0 & 0 \\ 0 & 0 & 0 & 0 & 1 & 0 & 0 & 0 \\ 0 & 0 & 0 & 0 & 0 & 0 & 0 & 0 \\ 0 & 0 & 0 & 0 & 0 & 0 & 0 & 0 \\ 0 & 0 & 0 & 0 & 0 & 0 & 0 & 0 \end{bmatrix} \in \mathbb{R}^{14 \times 8}, \tag{B.5}$$

$$C_1 = \begin{bmatrix} 0 & 0 & 0 & -1 & 0 & 0 & 0.015 & 0.015 \end{bmatrix} \in \mathbb{R}^{1 \times 8} \tag{B.6}$$

$$D_{00} = \begin{bmatrix}
0 & 1.6949 & 0 & 0 & 0 & 0 & 0 & 0 & 0 & 0 & 0 & 0 & 0 & 0 \\
0 & -1.6949 & 0 & 0 & 0 & 0 & 0 & 0 & 0 & 0 & 0 & 0 & 0 & 0 \\
0 & 0.2585 & -1.6949 & -1.6949 & 0 & 0 & 0 & 0 & 0 & 0 & 0 & 0 & 0 & 0 \\
0 & -1.6949 & 0 & 0 & 0 & 0 & 0 & 0 & 0 & 0 & 0 & 0 & 0 & 0 \\
-1 & 0.6949 & 0 & 0 & 0 & 0 & 0 & 0 & 0 & 0 & 0 & 0 & 0 & 0 \\
0 & 0 & 0 & 0 & 0 & 0 & 0 & 0 & 0 & 0 & 0 & 0 & 0 & 0 \\
0 & 0 & 0 & 0 & 0 & 0 & 1 & 0 & 0 & 0 & 0 & 0 & 0 & 0 \\
0 & 0 & 0 & 0 & 0 & 0 & 0 & 1 & 0 & 0 & 0 & 0 & 0 & 0 \\
0 & 0 & 0 & 0 & 0 & 0 & 0 & 0 & 1 & 0 & 0 & 0 & 0 & 0 \\
0 & 0 & 0 & 0 & 0 & 0 & -0.4891 & -0.0984 & 0.0608 & 0.0975 & 1 & 0 & 0 & 0 \\
0 & 0 & 0 & 0 & 0 & 0 & 0 & 0 & 0 & 0 & 0 & 0 & 0 & 0 \\
0 & 0 & 0 & 0 & 0 & 0 & 0 & 0 & 0 & 0 & 0 & 1 & 0 & 0 \\
0 & 0 & 0 & 0 & 0 & 0 & 0 & 0 & 0 & 0 & 0 & 0 & 1 & 0 \\
0 & 0 & 0 & 0 & 0 & 0 & 0 & 0 & 0 & 0 & 0 & 0 & 0 & 1
\end{bmatrix} \in \mathbb{R}^{14 \times 14},$$

$$(B.7)$$

$$
D_{01} =
\begin{bmatrix}
0 & 0 & 0.1161 & 1 \\
0 & 0 & -0.1161 & 0 \\
0 & 0 & 0.0073 & 0 \\
0 & 0 & -0.0476 & 0 \\
0 & 0 & 0.0476 & 0.41 \\
0 & 0 & 0 & 0 \\
0 & 0 & 0 & 0 \\
0 & 0 & 0 & 0 \\
0 & 0 & 0 & 0 \\
0 & 0 & 0 & 0 \\
0 & 0 & 0 & 0 \\
0 & 0 & 0 & 0 \\
0 & 0 & 0 & 0 \\
0 & 0 & 0 & 0
\end{bmatrix}
\in \mathbb{R}^{14 \times 3},
\tag{B.8}
$$

$$
D_{02} =
\begin{bmatrix}
1 \\
0 \\
0 \\
0 \\
0.41 \\
0 \\
0 \\
0 \\
0 \\
0 \\
0 \\
0 \\
0 \\
0
\end{bmatrix}
\in \mathbb{R}^{14 \times 1}
\tag{B.9}
$$

$$
D_{10} = [0\,0\,0\,0\,0\,0\,0\,0\,0\,0\,0\,0\,0\,0] \in \mathbb{R}^{1 \times 14},
\tag{B.10}
$$

$$
D_{10} = [0\,0\,0] \in \mathbb{R}^{1 \times 14},
\tag{B.11}
$$

$$
D_{10} = [0] \in \mathbb{R}.
\tag{B.12}
$$

Curriculum Vitae

Tamer Başar is with the University of Illinois at Urbana-Champaign, where he holds the academic positions of Swanlund Endowed Chair, Center for Advanced Study Professor of Electrical and Computer Engineering, Research Professor at the Coordinated Science Laboratory, and Research Professor at the Information Trust Institute. He received the B.S.E.E. degree from Robert College, Istanbul, and the M.S., M.Phil, and Ph.D. degrees from Yale University. He has published extensively in systems, control, communications, and dynamic games, and has current research interests that address fundamental issues in these areas along with applications such as formation in adversarial environments, network security, resilience in cyber-physical systems, and pricing in networks.

In addition to his editorial involvement with these Briefs, Basar is also the Editor-in-Chief of Automatica, Editor of two Birkhäuser Series on Systems & Control and Static & Dynamic Game Theory, the Managing Editor of the Annals of the International Society of Dynamic Games (ISDG), and member of editorial and advisory boards of several international journals in control, wireless networks, and applied mathematics. He has received several awards and recognitions over the years, among which are the Medal of Science of Turkey (1993); Bode Lecture Prize (2004) of IEEE CSS; Quazza Medal (2005) of IFAC; Bellman Control Heritage Award (2006) of AACC; and Isaacs Award (2010) of ISDG. He is a member of the US National Academy of Engineering, Fellow of IEEE and IFAC, Council Member of IFAC (2011–14), a past President of CSS, the Founding President of ISDG, and President of AACC (2010–11).

Antonio Bicchi is Professor of Automatic Control and Robotics at the University of Pisa. He graduated at the University of Bologna in 1988 and was a postdoc scholar at M.I.T. A.I. Lab between 1988 and 1990. His main research interests are in:

- Dynamics, kinematics, and control of complex mechanical systems, including robots, autonomous vehicles, and automotive systems;
- Haptics and dextrous manipulation; and
- Theory and control of nonlinear systems, in particular hybrid (logic/dynamic, symbol/signal) systems.

He has published more than 300 papers on international journals, books, and refereed conferences. Professor Bicchi currently serves as the Director of the Interdepartmental Research Center "E. Piaggio" of the University of Pisa, and President of the Italian Association or Researchers in Automatic Control. He has served as Editor in Chief of the Conference Editorial Board for the IEEE Robotics and Automation Society (RAS), and as Vice President of IEEE RAS, Distinguished Lecturer, and Editor for several scientific journals including the International Journal of Robotics Research, the IEEE Transactions on Robotics and Automation, and IEEE RAS Magazine. He has organized and co-chaired the first World Haptics Conference (2005), and Hybrid Systems: Computation and Control (2007). He is the recipient of several best paper awards at various conferences, and of an Advanced Grant from the European Research Council. Antonio Bicchi has been an IEEE Fellow since 2005.

Miroslav Krstic holds the Daniel L. Alspach chair and is the founding director of the Cymer Center for Control Systems and Dynamics at University of California, San Diego. He is a recipient of the PECASE, NSF Career, and ONR Young Investigator Awards, as well as the Axelby and Schuck Paper Prizes. Professor Krstic was the first recipient of the UCSD Research Award in the area of engineering and has held the Russell Severance Springer Distinguished Visiting Professorship at UC Berkeley and the Harold W. Sorenson Distinguished Professorship at UCSD. He is a Fellow of IEEE and IFAC. Professor Krstic serves as Senior Editor for Automatica and IEEE Transactions on Automatic Control and as Editor for the Springer series Communications and Control Engineering. He has served as Vice President for Technical Activities of the IEEE Control Systems Society. Krstic has co-authored eight books on adaptive, nonlinear, and stochastic control, extremum seeking, control of PDE systems including turbulent flows, and control of delay systems.

References

1. P. Apkarian, R.J. Adams, Advanced gain-scheduling techniques for uncertain systems. IEEE Trans. Control Syst. Technol. **6**(1), 21–32 (1998)
2. P. Apkarian, P. Gahinet, G. Becker, Self-scheduled $\mathscr{H}_\infty$ control of linear parameter-varying systems: a design example. Automatica **31**(9), 1251–1261 (1995)
3. P. Apkarian, P. Gahinet, A convex characterization of gain-scheduled $\mathscr{H}_\infty$ controllers. IEEE Trans. Autom. Control **40**(5), 853–864 (1995)
4. P. Apkarian, P.C. Pellanda, H.D. Tuan, Mixed H_2/H_∞ multi-channel linear parameter-varying control in discrete time. Syst. Control Lett. **41**(5), 333–346 (2000)
5. K.J. Astrom, B. Wittenmark, *Computer-Controlled Systems: Theory and Design* (Prentice Hall, New York, 1996)
6. G.J. Balas, J.C. Doyle, K. Glover, A. Packard, R. Smith, μ-analysis and synthesis toolbox. *Users Guide* (The MathWorks Inc., South Natick, 2001)
7. A. Balluchi, L. Benvenuti, M.D. di Benedetto, C. Pinello, A.L. Sangiovanni-Vincentelli, R. PARADES, Automotive engine control and hybrid systems: challenges and opportunities. Proc. IEEE **88**(7), 888–912 (2000)
8. K.A. Barbosas, C.E. de Souza, A. Trofino, Robust $\mathscr{H}_2$ filtering for discrete-time uncertain linear systems using parameter-dependent lyapunov functions, in *Proceedings of American Control Conference* (Anchorage, AK, May 2008)
9. N.F. Benninger, G. Plapp, Requirements and performance of engine management systems under transient conditions. *SAE 910083*, 1991
10. D.S. Bernstein, W.M. Haddad, LQG control with an h_∞ performance bound: a Riccati equation approach. IEEE Trans. Autom. Control **34**(3), 293–305 (1989)
11. J. Caigny, J. Camino, R. Oliveira, P. Peres, J. Swevers, Gain scheduled H_∞ -control of discrete-time polytopic time-varying systems, in *Proceedings of 47th IEEE Conference on Decision and Control*, 2008, pp. 3872–3877
12. J. Caigny, J. Camino, R. Oliveira, P. Peres, J. Swevers, Gain scheduled H_2 -control of discrete-time polytopic time-varying systems, in *Congresso Brasileiro de Automatica*, 2008
13. J. Caigny, J. Camino, R. Oliveira, P. Peres, J. Swevers, Gain scheduled $\mathscr{H}_2$ and $\mathscr{H}_\infty$ control of discrete-time polytopic time-varying systems. IET Control Theory Appl. **4**, 362–380 (2010)
14. J. Caigny, J. Camino, R. Oliveira, P. Peres, J. Swevers, Gain-scheduled dynamic output feedback control for discrete-time lpv systems. Int. J. Robust Nonlinear Control **22**, 535–558 (2012)
15. J. Cassidy Jr, M. Athans, On the design of electronic automotive engine controls using linear quadratic control theory. IEEE Trans. Autom. Control **25**(5), 901–912 (1980)
16. M. Chilali, P. Gahinet, $\mathscr{H}_\infty$ design with pole placement constraints: an LMI approach. IEEE Trans. Autom. Control **41**(3), 358–367 (1996)

A. P. White et al., *Linear Parameter-Varying Control for Engineering Applications*, 107
SpringerBriefs in Control, Automation and Robotics,
DOI: 10.1007/978-1-4471-5040-4, © The Author(s) 2013

17. S.B. Choi, J.K. Hedrick, V. Kelsey-Hayes, M.I. Livonia, An observer-based controller design method for improving air/fuel characteristics of spark ignition engines. IEEE Trans. Control Syst. Technol. **6**(3), 325–334 (1998)

18. M.C. De Oliveira, J.C. Geromel, J. Bernussou, Extended H_2 and H_∞ norm characterizations and controller parametrizations for discrete-time systems. Int. J. Control **75**(9), 666–679 (2002)

19. M.C. de Oliveira, J. Bernussou, J.C. Geromel, A new discrete-time robust stability condition. Syst. Control Lett. **37**(4), 261–265 (1999)

20. M.C. De Oliveira, J.C. Geromel, J. Bernussou, An LMI optimization approach to multiobjective controller design for discrete-time systems, in *Proceedings of the 38th IEEE Conference on Decision and Control*, vol. 4 (IEEE, 1999), pp. 3611–3616

21. P.H. Dugdale, R.J. Rademacher, B.R. Price, J.W. Subhedar, R.L. Duguay, Ecotec 2.4l VVT: a variant of GM's global 4-cylinder engine. *SAE 2005–01-1941*, 2005

22. P. Gahinet, Explicit controller formulas for LMI-based H_∞ synthesis. Automatica **32**(7), 1007–1014 (1996)

23. P. Gahinet, A. Nemirovski, A.J. Laub, M. Chilali, *Matlab LMI Control Toolbox* (The MathWorks Inc, Natick, 1995)

24. P. Gahinet, P. Apkarian, M. Chilali, Affine parameter-dependent Lyapunov functions and real parametric uncertainty. IEEE Trans. Autom. Control **41**(3), 436–442 (1996)

25. A.U. Genç, K. Glover, R. Ford, Nonlinear control of hydraulic actuators in variable cam timing engines, in *MECA International Workshop* (University of Salerno, Italy, Sept 2001)

26. A.U. Genç, Linear parameter-varying modelling and robust control of variable cam timing engines. PhD thesis, University of Cambridge, 2002

27. M. Green, D.J.N. Limebeer, *Linear Robust Control* (Prentice-Hall, New Jersey, 1995)

28. G.G. Zhu, R.E. Skelton, L_2 to L_∞ gains for sampled-data systems. Int. J. Control **61**(1), 19–32 (1995)

29. L. Guzzella, C.H. Onder, *Introduction to Modeling and Control of Internal Combustion Engine Systems* (Springer, Berlin, 2004)

30. J.B. Heywood, D.R. Cohn, L. Bromberg, Optimized fuel management system for direct injection ethanol enhancement of gasoline engines. US Patent 7,225,787, 5 June 2007

31. T. Ikoma, S. Abe, Y. Sonoda, H. Suzuki, Y. Suzuki, M. Basaki, Development of v-6 3.5-liter engine adopting new direct injection system. *SAE 2006–01-1259*, 2016:293, 2006

32. I. Kaminer, P.P. Khargonekar, M.A. Rotea, Mixed h_2/h_∞ control for discrete-time systems via convex optimization. Automatica **29**(1), 57–70 (1993)

33. P.P. Khargonekar, M.A. Rotea, Mixed h_2/h_∞ control: a convex optimization approach. IEEE Trans. Autom. Control **36**(7), 824–837 (1991)

34. M. Kočvara, M. Stingl, Penbmi user's guide (version 2.1), 2006

35. J. Löfberg, Yalmip: a toolbox for modeling and optimization in MATLAB, in *Proceedings of the CACSD Conference* (Taipei, Taiwan, 2004)

36. I. Masubuchi, A. Ohara, N. Suda, LMI-based controller synthesis: a unified formulation and solution. Int. J. Robust Nonlinear Control **8**(8), 669–686 (1998)

37. L. Mianzo, H. Peng, I. Haskara, Transient air-fuel ratio h_∞ preview control of a drive-by-wire internal combustion engine, in *Proceedings of American Control Conference*, 2001, pp. 2867–2871

38. Y. Moriya, A. Watanabe, H. Uda, H. Kawamura, M. Yoshiuka, M. Adachi, A newly developed intelligent variable valve timing system–continuously controlled cam phasing as applied to new 3 liter inline 6 engine. *SAE 960579*, 1996

39. K.R. Muske, J.C.P. Jones, A model-based SI engine air fuel ratio controller, in *Proceedings of American Control Conference*, 2006, pp. 3284–3289

40. R. Nagamune, J. Choi, Parameter reduction of estimated model sets for robust control. J. Dyn. Syst. Meas. Control **132**(2), 10 (2010). doi:10.1115/1.4000661

41. R.C.L.F. Oliveira, P.L.D. Peres, Robust stability analysis and control design for time-varying discrete-time polytopic systems with bounded parameter variation, in *Proceedings of 2008 American Control Conference* (Seattle, WA, June 2011)

42. R.C.L.F. Oliveira, P.L.D. Peres, Time-varying discrete-time linear systems with bounded rates of variation: stability analysis and control design. Automatica **45**(11), 2620–2626 (2009)

43. C.H. Onder, Model-based multivariable speed and air-to-fuel ratio control of an SI engine. *SAE 930859*, 1993

44. S. Pace, G.G. Zhu, Sliding mode control of a dual-fuel system internal combustion engine, in *Proceedings of ASME Dynamic Systems and Control Conference* (Hollywood, CA, Oct 2009)

45. A. Packard, Gain scheduling via linear fractional transformations. Syst. Control Lett. **22**(1), 79–92 (1994)

46. J.D. Powell, N.P. Fekete, C.F. Chang, Observer-based air-fuel ratio control. IEEE Control Syst. Mag. **18**, 72–83 (1998)

47. W.F. Powers, B.K. Powell, G.P. Lawson, Applications of optimal control and Kalman filtering to automotive systems. Int. J. Veh. Des. *SP4*, 39–53 (1983)

48. Z. Ren, G. Zhu, Pseudo-random binary sequence closed-loop system identification error with integration control. Proc. Inst. Mech. Eng. Part I J. Syst. Control Eng. **223**(6), 877–884 (2009) doi:10.1243/09596518JSCE794

49. Z. Ren, G.G. Zhu, Integrated system ID and control design for an IC engine variable valve timing system. ASME J. Dyn. Syst. Meas. Control (2010) doi:10.1115/1.4003263

50. J.G. Rivard, Closed-loop electronic fuel injection control of the internal-combustion engine. *SAE 730005*, 1973

51. A. Rotea, P.P. Khargonekar, h^2-optimal control with an h^∞-constraint: the state feedback case. Automatica **27**(2), 307–316 (1991)

52. C. Scherer, P. Gahinet, M. Chilali, Multiobjective output-feedback control via LMI optimization. IEEE Trans. Autom. Control **42**(7), 896–911 (1997)

53. S.J. Shamma, M. Athans, Guaranteed properties of gain scheduled control for linear parameter-varying plants. Automatica **27**(3), 559–564 (1991)

54. S. Skogestad, I. Postlethwaite, *Multivariable Feedback Control: Analysis and Design*, 2nd edn. (Wiley, New York, 2005)

55. A.A. Stoorvogel, The robust $\mathscr{H}_2$ control problem: a worst-case design. IEEE Trans. Autom. Control **38**(9), 1358–1370 (1993)

56. J.F. Sturm, Using SeDuMi 1.02, a MATLAB toolbox for optimization over symmetric cones. Optim. Methods Softw. **11**(1), 625–653 (1999)

57. K. Suzuki, T. Shen, J. Kako, Y. Oguri, Individual A/F control with fuel-gas ratio estimation for multi-cylinder IC engines, in *Proceedings of American Control Conference*, 2007, pp. 5094–5099

58. L.N. Trefethen, D. Bau, Numerical linear algebra. Soc. Ind. Appl. Math. **39**, 575–809 (1997)

59. R.C. Turin, H.P. Geering, Model-reference adaptive A/F ratio control in an SI engine based on Kalman-filtering techniques, in *Proceedings of American Control Conference*, 1996, pp. 4082–4090

60. J. Warren, S. Schaefer, A.N. Hirani, M. Desbrun, Barycentric coordinates for convex sets. Adv. Comput. Math. **27**(3), 319–338 (2007)

61. A. White, J. Choi, R. Nagamune, G. Zhu, Gain-scheduling control of port-fuel-injection processes. IFAC J. Control Eng. Pract. **19**(4), 380–394 (2011). doi:10.1016/j.conengprac.2010.12.007

62. A. White, G. Zhu, J. Choi, Hardware-in-the-loop simulation of robust gain-scheduling control of port-fuel-injection processes. IEEE Trans. Control Syst. Technol. **19**(6), 1433–1443 (2011). doi:10.1109/TCST.2010.2095420

63. A. White, G. Zhu, J. Choi, Mixed $\mathscr{H}_2/\mathscr{H}_\infty$ observer-based LPV control of a hydraulic engine cam phasing actuator. IEEE Trans. Control Syst. Technol. **21**(1), 229–238 (2013). doi:10.1109/TCST.2011.2177464

64. D.A. David, Convolution and hankel operator norms for linear systems. IEEE Trans. Autom. Control **34**(1), 94–97 (1989)
65. F. Wu, K. Dong, Gain-scheduling control of LFT systems using parameter-dependent Lyapunov functions. Automatica **42**(1), 39–50 (2006)
66. F. Wu, A generalized LPV system analysis and control synthesis framework. Int. J. Control **74**(7), 745–759 (2001)
67. F. Wu, X.H. Yang, A. Packard, G. Becker, Induced l_2-norm control for LPV systems with bounded parameter variation rates. Int. J. Robust Nonlinear Control **6**, 2379–2383 (1996)
68. X. Yang, G.G. Zhu, A mixed mean-value and crank-based model of a dual-stage turbocharged SI engine for hardware-in-the-loop simulation, in *Proceedings of American Control Conference* (Baltimore, MD, June, 2010), p. 2010
69. Y. Yildiz, A. Annaswamy, D. Yanakiev, I. Kolmanovsky, Adaptive air fuel ratio control for internal combustion engines, in *Proceedings of American Control Conference*, 2008, pp. 2058–2063
70. J. Yu, A. Sideris, $\mathscr{H}_\infty$ control with parametric Lyapunov functions. Syst. Control Lett. **30**(1), 57–69 (1997)
71. F. Zhang, K.M. Grigoriadis, M.A. Franchek, I.H. Makki, Linear parameter-varying lean burn air-fuel ratio control for a spark ignition engine. J. Dyn. Syst. Meas. Control **192**, 404–414 (2007)
72. K. Zhou, J.C. Doyle, *Essentials of Robust Control* (Prentice Hall, Upper Saddle River, 1998)
73. K. Zhou, J.C. Doyle, K. Glover, *Robust and Optimal Control* (Prentice Hall, Upper Saddle River, 1996)
74. G. Zhu, $\mathscr{L}_2$ and $\mathscr{L}_\infty$ Multiobjective control for linear systems. Ph.D. thesis, Purdue University, 1992
75. G. Zhu, K.M. Grigoriadis, R.E. Skelton, Covariance control design for hubble space telescope. J. Guid. Control Dyn. **18**(2), 230–236 (1995)
76. G. Zhu, M. Rotea, R.E. Skelton, A convergent algorithm for the output covariance constraint control problem. SIAM J. Control Optim. **35**, 341–361 (1997)
77. R.A. Zope, J. Mohammadpour, K.M. Grigoriadis, M. Franchek, Air-fuel ratio control of spark ignition engines with TWC using LPV techniques, in *Proceedings of ASME Dynamic Systems and Control Conference*, 2009